Conservation Research 1995

STUDIES IN THE HISTORY OF ART • 51 •

Monograph Series II

Conservation Research 1995

National Gallery of Art, Washington

Distributed by the University Press of New England
Hanover and London

This publication was produced by the Editors Office, National Gallery of Art, Washington
Editor-in-Chief, Frances P. Smyth

The type is Trump Medieval, set by Artech Graphics II, Inc., Baltimore, Maryland

The text paper is 128 gsm Japanese matt

Printed in Hong Kong

Distributed by the University Press of New England, 23 South Main Street, Hanover, New Hampshire 03755

Abstracted and indexed in BHA (Bibliography in the History of Art) and Art Index

Conservation Research is a periodic publication of the National Gallery of Art, Washington. The primary objective of *Conservation Research* is to report regularly on the research and technical studies pursued by the conservation division of the National Gallery of Art

ISSN 0091-7338
ISBN 0-89468-208-3

Cover: David Smith, *Circle III*, 1962, painted steel, National Gallery of Art, Washington, Ailsa Mellon Bruce Fund

Contents

Preface

The traditional goals of conservation have been to preserve works of art by maintaining their structural security and appearance and by controlling the environments in which they are exhibited and stored. Scientific analyses have enhanced conservators' understanding of the construction and condition of works of art, thereby enabling them to make sound decisions concerning treatment and care. While public attention is often drawn to the conservator's ability to date materials, thus at times indicating that a work of art may be misattributed, this focus constitutes only a small part of analytical work. In addition to the highly developed manual abilities conservators must possess, they have become skilled at relating analytical data to a range of art-historically significant conclusions. Increasing emphasis is also placed on the application of scientific findings to conservation methodology.

Therefore, the best conservation results from a collaborative effort by the art historian, conservation scientist, and conservator. The conservator today is charged with understanding the ever-changing technologies as new instruments and tools become applicable. For example, a unique infrared camera (given to the National Gallery by Eastman Kodak) has allowed us to gain more insight into artists' working techniques and has established computer image processing as an important tool. The utilization of computers and imaging technology and the application of sophisticated scientific procedures such as high-performance liquid chromatography (HPLC) and x-ray fluorescence spectrometry (XRF), commonly used in other disciplines, have become useful tools for the conservation field, as several articles in this volume attest. The use of facsimiles and the establishment of a protocol for XRF permit identification of platinum and palladium finishing processes in the photographs by Alfred Stieglitz. A method of determining the presence of amino acids in artists' materials dating from the Renaissance to the late nineteenth century has been developed using HPLC.

In this volume also, a study of David Smith's working methods enables us to establish the paint finishes and the extent of layering Smith used in a number of his sculptures. Why tideline formation in paper objects occurs, the resulting inherent degradation processes, occurrence and reoccurrence, and methods of alleviating the condition are explored by means of a rigorous testing procedure. Finally, the traditions, literature, and working methods of print restoration are investigated.

The rapid development of new technical information about works of art and its impact on scholarship require a wide distribution of the results so that these insights may be not only shared but applied and expanded. Our publishing goals, therefore, are to encourage application of new technologies to the study of works of art, to maintain a high standard of practical treatment, and, with our curatorial colleagues, to integrate our knowledge.

Acknowledgments

We are indebted to Earl A. Powell III, director of the National Gallery of Art, as well as his predecessor, J. Carter Brown, for supporting conservation projects and this publication. We also thank the critical readers who have taken time from their own endeavors to review the manuscripts; the staff and fellows who have researched and prepared these essays; and the department heads who have read the articles and offered their assistance. In addition we thank Janice Gruver, conservation division editor, for organizing and coordinating the process.

The generous support of the Conservation Fellowship Endowment sponsors, principally the Andrew W. Mellon Foundation, as well as the Charles E. Culpeper Foundation, has provided our fellows with the opportunity to undertake research projects. We are also indebted to the Publications Fund of the National Gallery of Art for support.

ROSS M. MERRILL
Chief of Conservation, National Gallery of Art

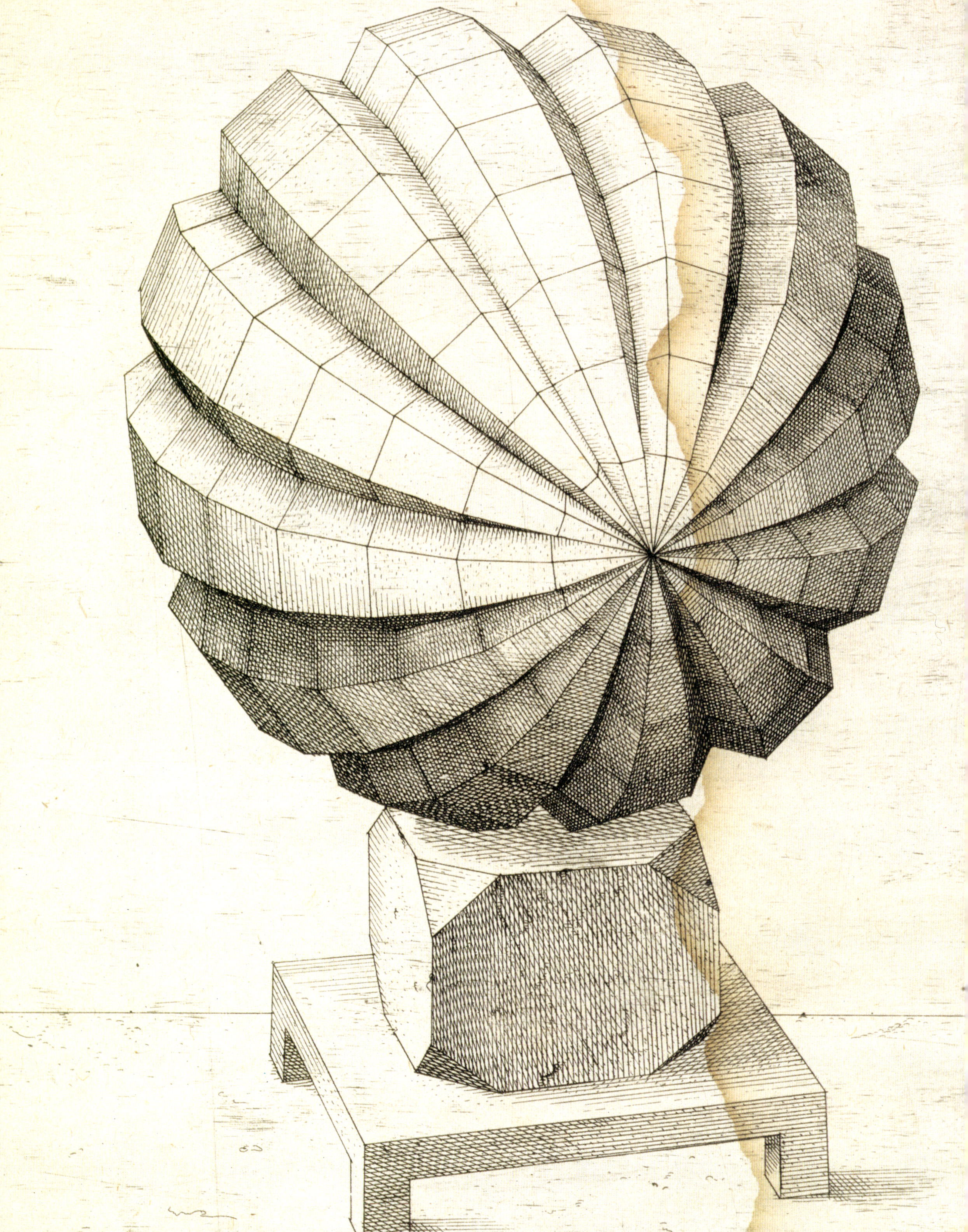

ELMER EUSMAN

Tideline Formation in Paper Objects: Cellulose Degradation at the Wet-Dry Boundary

The vital, yet deadly, affinity between cellulose and water thus represents a true conflict of interests, and to find the optimum balance remains a continuing challenge.

D. Heinz Corte, *Handbook of Paper Science*

The growing concern among paper conservators over possible side effects of local treatment and the recent suggestions of a possible link between evaporating liquid systems and foxing[1] were the impetus for this research.

Brown lines resulting from water damage are a familiar problem facing the paper conservator (fig. 1a–b). But, depending on the type of paper and the liquid used, local treatment can also lead to stain formation around the treated area, manifested as a faint yellow to deep brown discolored line. The application of an aqueous solution during the treatment of an eighteenth-century paper, for example, often produces colored compounds at the wet-dry boundary, commonly believed to be the result of the migration and concentration of water-soluble compounds. This assumption is reinforced by the lighter color of the wetted region of the paper. Under ultraviolet light, this discoloration is revealed as regularly arising in combination with an increased fluorescence in the area of the wet-dry boundary. While treating a more modern paper generally does not produce discoloration around the treated area, increased fluorescence can be observed under ultraviolet light. In addition to water and aqueous solutions, organic solvents also have the capability of creating fluorescent rings around the treated area. This enhanced fluorescence has generally been attributed to the resins present in paper and their deposit at the boundary area of the wet-dry interface. Although more research into the effects produced by an evaporating organic solvent is needed, this study addresses primarily aqueous solutions.

Another aspect of the wet-dry boundary area was described for the textile industry in 1934 by W. H. Bone.[2] Not only did he demonstrate that water was capable of creating brown tidelines in highly purified cotton cellulose cloth, but he also noted the presence of a cellulose modification not visible in either day or ultraviolet light. Bone's research was the first indication that processes other than migration and concentration of water-soluble products were taking place in the formation of water stains. The first section of this essay will discuss the nature of these processes.

Conservators refer to the brown stains formed at the liquid-dry boundary in paper as "tidelines"; this term will be used to describe the stains that form after paper is locally wetted by water, an aqueous solution, or organic solvent. Fluorescent tidelines, which are visible only in ultraviolet light and are caused by aqueous solutions or organic solvents, are of the most concern to conservators. The suggestion that these lines appear over time as brown lines[3] has prompted practicing conservators to inquire about the relationship between the fluorescent and the colored com-

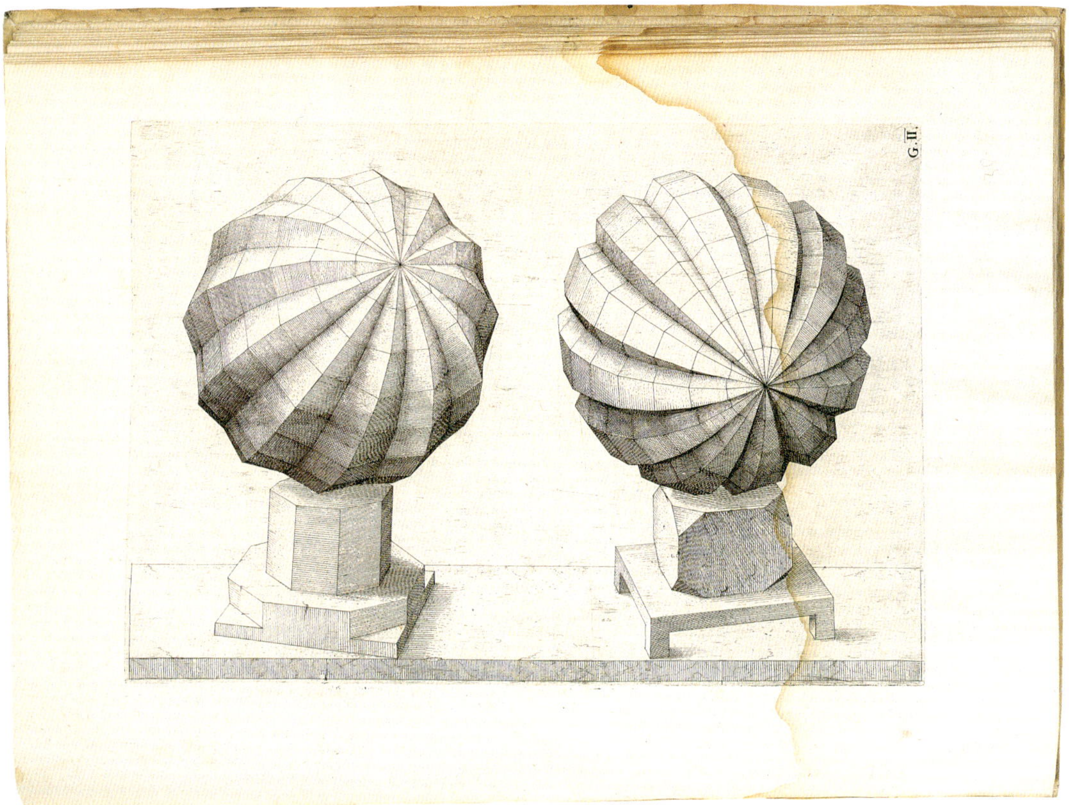

1a. Wenzel Jamnitzer I, *Perspectiva Corporum Regularium*, bound volume (1568). Discoloration was caused by water penetrating the book at the bottom
National Gallery of Art, Washington, Mark J. Millard Architectural Collection

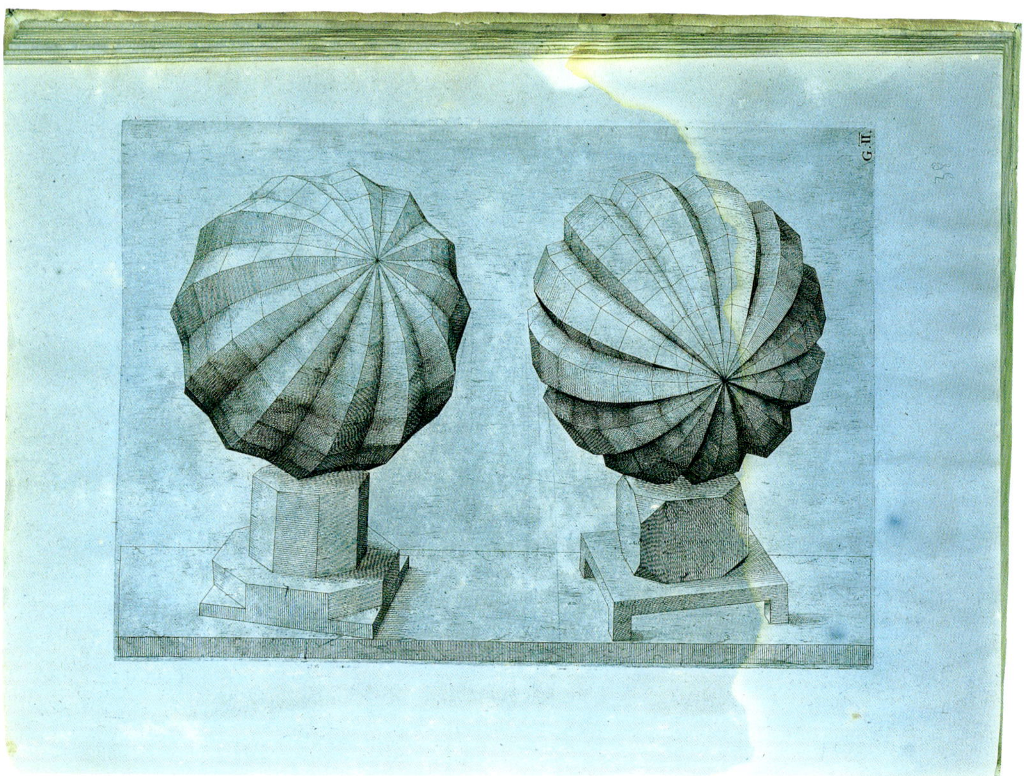

1b. Photograph of fig. 1a under ultraviolet light. Note the increased fluorescence in the area of the tideline

pounds. To answer this question, the second part of this study addresses the influence of different papers, liquids, and methods of application on the formation of a tideline.

Tests Used to Observe Tideline Discoloration

A simple experiment to create tidelines deliberately was described in textile industry literature between 1950 and 1965.[4] Strips of cotton cloth or paper were suspended vertically with the lower end dipped in water, leaving the other end exposed to air. The liquid rose owing to capillary action until it reached a point at which the steady rise was balanced by evaporation from the wetted surface. As soon as the liquid reached its highest position, the tideline region could be detected by an increased fluorescence, and in a matter of hours a brown line formed at the boundary between the wet and dry region.

Cellulose degradation in the tideline region was first demonstrated by Bone using methylene blue. This blue dye is an excellent tool for disclosing the presence of carboxylic groups in cellulosic materials.[5] Immersing a strip sample of Whatman number 1 (hereafter Whatman 1) filter paper containing a tideline in a solution of methylene blue showed a clear increased absorption in the tideline region. Similar tests confirming these results were conducted in the paper conservation department of the National Gallery of Art (fig. 2a–c).[6] Bone and H. A. Turner noted that generally the yellow-brown discoloration and the fluorescence could be washed out immediately after the sample had dried at room temperature.[7] However, the tideline region in the washed sample still showed the same degree of methylene blue absorption as before washing. This same phenomenon was also observed in the tests conducted at the National Gallery of Art, implying that in addition to the formation of water-soluble colored products, water-insoluble products containing carboxylic groups are created. Further indications of cellulose degradation, reported by Bone and others, included an enhanced absorption of Harrison's solution,[8] an increase in cuprammonium fluidity,[9] and a severe tendering of the cloth in the tideline region.[10]

The literature provides several views concerning the mechanism of tideline formation that should be considered by paper conservators when defining the nature of tideline formation:

- Evaporation of distilled water from cotton cellulose, under the conditions described above, results in a sharp, well-defined boundary area between the wet and dry region. This boundary area contains two identifiable cellulose degradation products: a yellow-brown and fluorescent product that can be extracted in water, leaving no trace of discoloration or fluorescence in the cellulose, and a non-water-soluble product that, after extracting the yellow-brown fluorescent product, accounts for the increased methylene blue absorption.
- If the setup is amended by progressively raising the water level, causing the wet-dry boundary on the strip to overlap previous tidelines, only the top line shows brown discoloration and fluorescence. The water-soluble brown-colored fluorescent material is transported to the last and highest wet-dry boundary, leaving no colored products and only very faintly fluorescent products at the previous tidelines. However, the previous boundary areas still show an increased methylene blue absorption, demonstrating that degradation of cellulose takes place at every wet-dry boundary, although the discoloration and fluorescence have been washed out. Progressively lowering the water level results in several discolored lines wherever a steady wet-dry boundary existed for a number of hours. Apart from a darker first line, all lines are equally discolored and all show fluorescence and an enhanced methylene blue absorption. Apparently, new colored compounds are created at every new level, even when all the present water-soluble compounds are transported out of the wet region. Tests at the National Gallery of Art described above using Whatman 1 filter paper produced similar results (figs. 3a–c and 4a–c).

Other factors in tideline formation have been investigated. The influence of oxygen on stain formation was tested by G. W. Madaras and H. A. Turner.[11] Robert Schaffer, William D. Appel, and Florence H. Forziati conducted experiments with various fibrous materials and solutions.[12] M. R. Fox suggested chemical groups in cellulose that could be responsible for the discoloration in the tideline region.[13] Of particular interest to conservators are some observations noted in the literature:

- Tidelines form not only on commercially prepared (bleached or unbleached) cloths, even when pretreated with deionized water, but

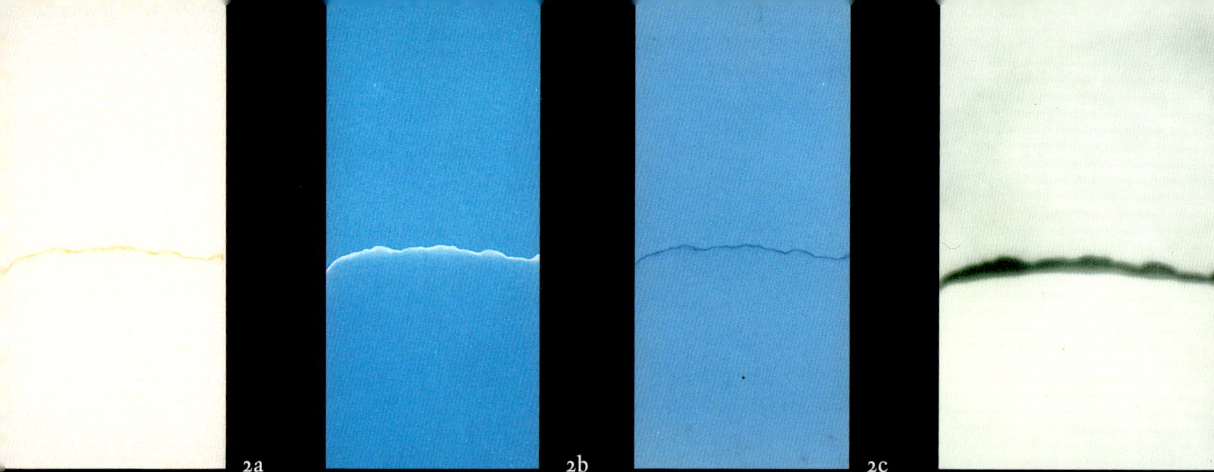

2a 2b 2c 2d

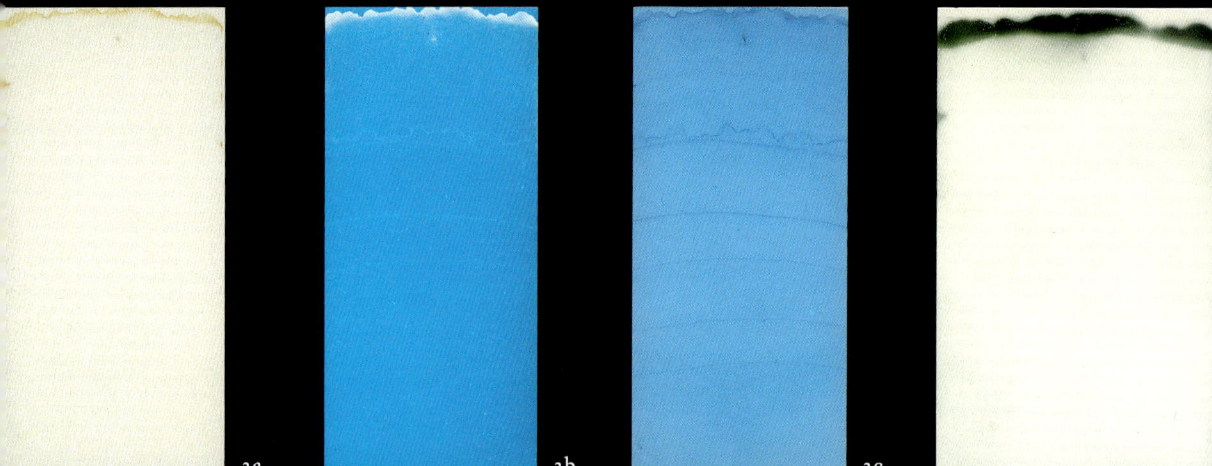

3a 3b 3c 3d

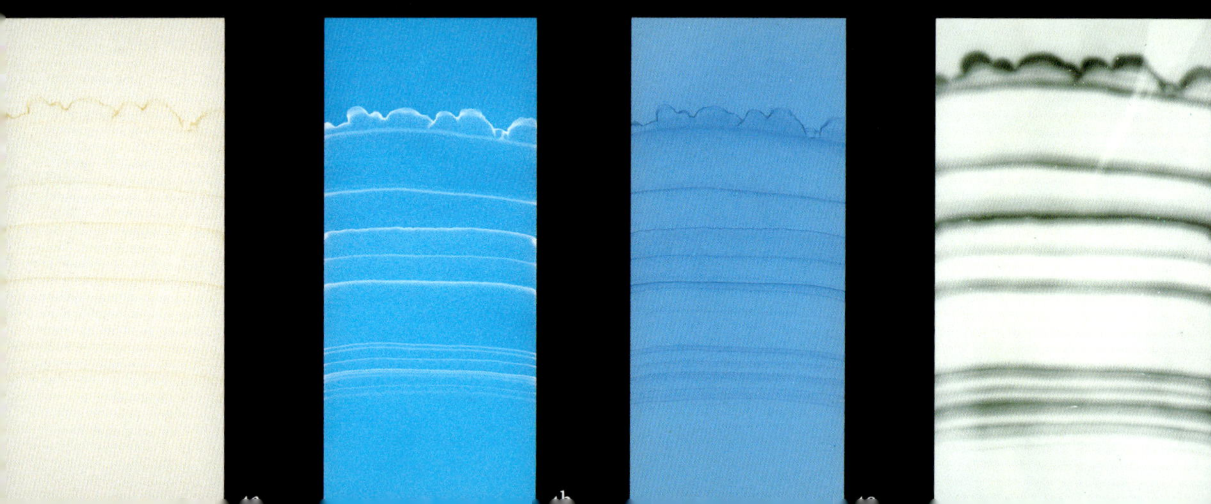

2a. The tideline coloration shown in a test strip of Whatman 1 filter paper was created over a period of two days

2b. Increased fluorescence of the tideline is visible viewed under ultraviolet light

2c. The tideline is visible through an increased absorption of methylene blue

2d. A Russell-effect image of the stained area shows the increased concentration of peroxides in the tideline

3a. The tideline coloration in a test strip of Whatman 1 filter paper was created by raising the wet-dry boundary in several increments over the course of three days

3b. Increased fluorescence of the tidelines is visible viewed under ultraviolet light

3c. The tidelines are visible through an increased absorption of methylene blue

3d. A Russell-effect image of the stained area shows the increased concentration of peroxides in the tideline

4a. The tideline coloration shown in a test strip of Whatman 1 filter paper was created by lowering the wet-dry boundary in several increments over the course of three days

4b. Increased fluorescence of the tidelines is visible viewed under ultraviolet light

4c. The tidelines are visible through an increased absorption of methylene blue

4d. A Russell-effect image of the stained area shows the increased concentration of peroxides in the tidelines

also on highly purified cloths. They form when distilled, deionized, hard, and soft water are used and also when dilute solutions of common salt, hydrochloric acid, sodium carbonate, and sodium hydroxide are used.[14]

- Tests conducted using an acetic acid solution (pH 3.9) and an ammonium hydroxide solution (pH 10.5) reveal that the increase in methylene blue absorption is larger for the ammonium hydroxide sample.[15]
- Fox showed that the tideline experiment results in a brown line in sunlight, diffuse daylight, artificial light, and total darkness, a result that appears to discount photochemical reactions as a major contributor to tideline formation.[16]
- The formation of a brown line does not seem to be related to biological activity since tideline formation occurs in fabrics that are pretreated with solutions of antiseptic agents such as thymol. However, bringing the fungus *Aspergillus niger* in contact with a sample containing a tideline area and storing the sample at 30°C resulted in rapid growth along the brown line. It took five days for the fungus to spread to other areas.[17]
- The formation of a tideline is not inhibited when the test is conducted with chelating agents such as oxalic acid.[18]
- The brown, fluorescent product is soluble in water when the experiment is conducted at room temperature and the sample is not dried at high temperatures.[19]
- Several cases show that water can form a wet-dry boundary lengthwise, widthwise, and crosswise in cloth. It has been suggested, therefore, that water can also form an active boundary near the face or the back of a fabric through the *thickness* of the cloth.[20]

The repeated discoloration and development of fluorescence only at the tideline rather than throughout the wetted area suggest a very active boundary between the wet and dry portions of the sheet. Some researchers speculated on the cause of the degradation in the tideline region. Bone and Turner proposed that minute shifts in the position of the tideline allow better penetration of oxygen to the tideline region, which is never completely wet or dry but contains swollen fibers. The wetted fibers would be protected from oxygen attack by a layer of water. Bone and Turner also speculated that the continuous wetting and drying of fibers in the small area of the tideline fluctuation might cause a disruption of the cellular chain, thus allowing increased chemical attack.[21]

Vertically Suspended Sample Experiments

The formation of discoloration at the tideline region seems to be a dual process. Migration of water-soluble products and their concentration at the wet-dry boundary appear to occur; however, analysis also established that a stationary degradation process takes place in the same region. Both aspects of the process were investigated. Although it was suspected that the damage to the cellulose accompanying the tideline formation was caused by oxidative processes, this connection had not been proven in earlier experiments. To verify that oxidation had taken place at the tideline, several experiments were conducted.

Relative Peroxide Content

Because the presence of peroxides in organic material is a clear indication of oxidation reactions, tidelines in Whatman 1 filter paper were tested for peroxides. Peroxide concentration measurements were taken by Suzanne Quillen Lomax, organic chemist at the National Gallery of Art.[22] The method of analysis consisted of reacting iodide ion with hydroperoxides, peroxides, and peracids to form triiodide ion. The concentration of the triiodide can be measured spectrophotometrically at 362 nanometers.

Measurements were taken directly from the paper and from water extractions. Paper samples or water extracts were placed in an isopropanol-glacial acetic acid solution of sodium iodide. After deaeration and dark storage, the concentration of triiodide was determined. The difference between the amounts of peroxides in the paper and the water-extracted peroxides was very small; a large part of the peroxides present in the tideline region was water soluble. The peroxide content in the tideline region was repeatedly found to be ten to twelve times as high as in the wet or dry region (fig. 5). It was also found that the region above the tideline generally contained a larger amount of peroxides than the region below the tideline. Part of the higher concentration of peroxides was evidently created by transportation to the tideline by the rising and evaporating water.

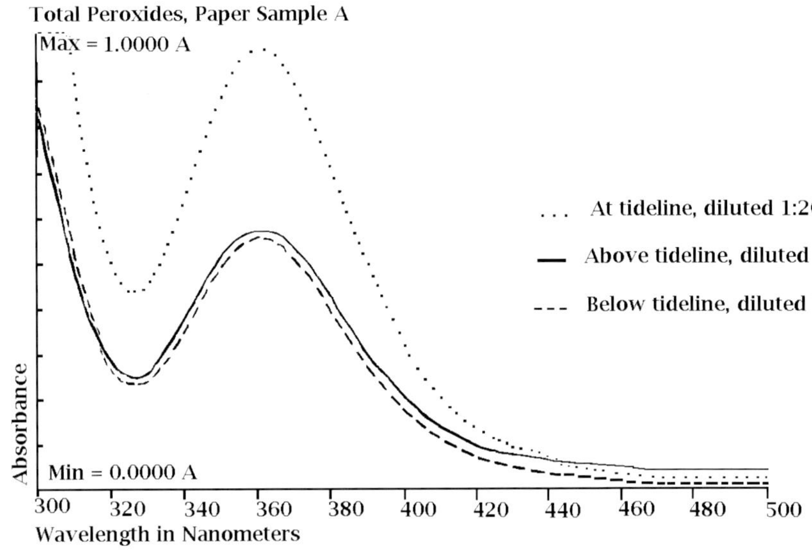

5. Ultraviolet-visible spectrum showing absorbance maximum for peroxide concentration in unwashed Whatman 1 filter paper. Maximum concentration is at 362 nm in the region of the tideline. Note that the extraction made from the tideline region was diluted by a factor 10

However, the difference in peroxide concentration between the region above and below the tideline was often very small and in some cases nonexistent.

The high concentration of peroxides at the tideline could be explained by their catalytic effect. The presence of peroxides is a strong indication of an autoxidation process. Peroxides are not only the products from this chain reaction; they also function as initiators. Peroxides are unstable and produce free radicals that react with oxygen to form new peroxides. The high concentration of peroxides in the tideline therefore appears to be the result of a combination of transportation and concentration and the subsequent chain reaction that peroxides initiate.

Russell-Effect Photography

Information about Russell-effect photography has been recently published.[23] In short, the procedure for producing a Russell-effect image consists of sensitizing a silver gelatin film[24] with an ammonium hydroxide solution and bringing the dried film in direct contact with the sample for twenty-four hours.[25] Sensitizing, exposure, and development of the film must take place in total darkness. Although the exact mechanism of image formation using this technique remains unclear, it is assumed that peroxides cause the silver ions to be reduced to silver metal, as does light during the normal exposure of a film. The suggestion that peroxides are responsible for this effect is based on circumstantial evidence supported by the results of several experiments.[26]

Good results were achieved using Guilleminot EG 18 film.[27] The images showed the tideline as a dark area on the film, indicating the higher concentration of peroxides in this region. Russell-effect images of tidelines were obtained in earlier research,[28] but these images had not been combined with a definite analysis of a higher peroxide concentration in the same area. In figure 2d (Russell-effect photograph of fig. 2a), the region above the tideline shows more patches of gray in the film than that below the tideline. This result further suggests that peroxides, present in paper, are transported to the wet-dry boundary during a tideline test. That the peroxides are largely water-soluble is clearly visible in figure 3d (Russell-effect photograph of fig. 3a). All the peroxides have been transported to the top tideline, leaving no traces of previous wet-dry boundaries. The generation of new peroxides at every tideline is shown in figure 4d (Russell-effect photograph of fig. 4a). Lowering the water level in several increments resulted in multiple lines with high concentrations of peroxides at former wet-dry boundaries.

A number of experiments were conducted to investigate the influence of antioxidants on the tideline formation process. Unfortunately, the results were inconclusive.[29]

Water-Soluble Degradation Products

An experiment was designed to determine the importance of migration of water-soluble compounds already present in the paper sample in the formation of a tideline. Whatman 1 filter paper samples were washed for three hours in high-performance liquid chromatography (HPLC)-grade water. The water was changed three times to remove all water-soluble material from the sheets. The paper strips were then air-dried for twenty-four hours. After all the strips were dried, they were placed in dark storage in a museum environment. Each strip was subjected to a tideline formation test over a period of several weeks.

The first test was conducted immediately after the strips were washed and air-dried. The unwashed strips were used as a control. The washed and unwashed paper samples were exposed to the tideline test for twenty-four hours. The test showed that much less discoloration occurred in the tideline area in the washed sample than in the unwashed control sample. This finding indicates that a large amount of the material responsible for the discoloration was washed out of the paper (fig. 6a, left: unwashed Whatman 1 filter paper; center: washed Whatman 1 filter paper).

However, a sample that was stored for two weeks showed a fainter tideline than was observed in the unaged unwashed control, but a darker tideline than was observed in the washed paper sample. Storage in an environment with no light for one month was sufficient to form a tideline as dark as the unwashed control sample (fig. 6a, right). It appears that the water-soluble compounds that were washed from the sheets can reform in the sheets after a short time and that these compounds are paper degradation products that may add to discoloration in the tideline.

The differences in fluorescence and methylene blue absorption are small (figs. 6b–c). The Russell-effect image of the unwashed and washed samples shows small differences in the peroxide concentration in the tideline region (fig. 6d). This result supports the assumption of peroxide formation in the tideline. The gray patches in the top part of the washed and stored test strip indicate peroxide formation in the entire strip during storage.

If migration of the degradation products and their concentration at the wet-dry boundary is part of the mechanism that creates a tideline, the following observations by J. Strachan as well as G. W. Huffman and colleagues are of interest. Although cellulose is insoluble in water, it is possible to

6a. Unwashed (left) and washed (center) and unwashed control (right) Whatman 1 filter paper test strips showing tideline formation and coloration

6b. Increased fluorescence of the tidelines are visible viewed under ultraviolet light

6c. The tidelines are visible through an increased absorption of methylene blue

6d. A Russell-effect image detail of the stained area shows the increased concentration of peroxides in the tidelines

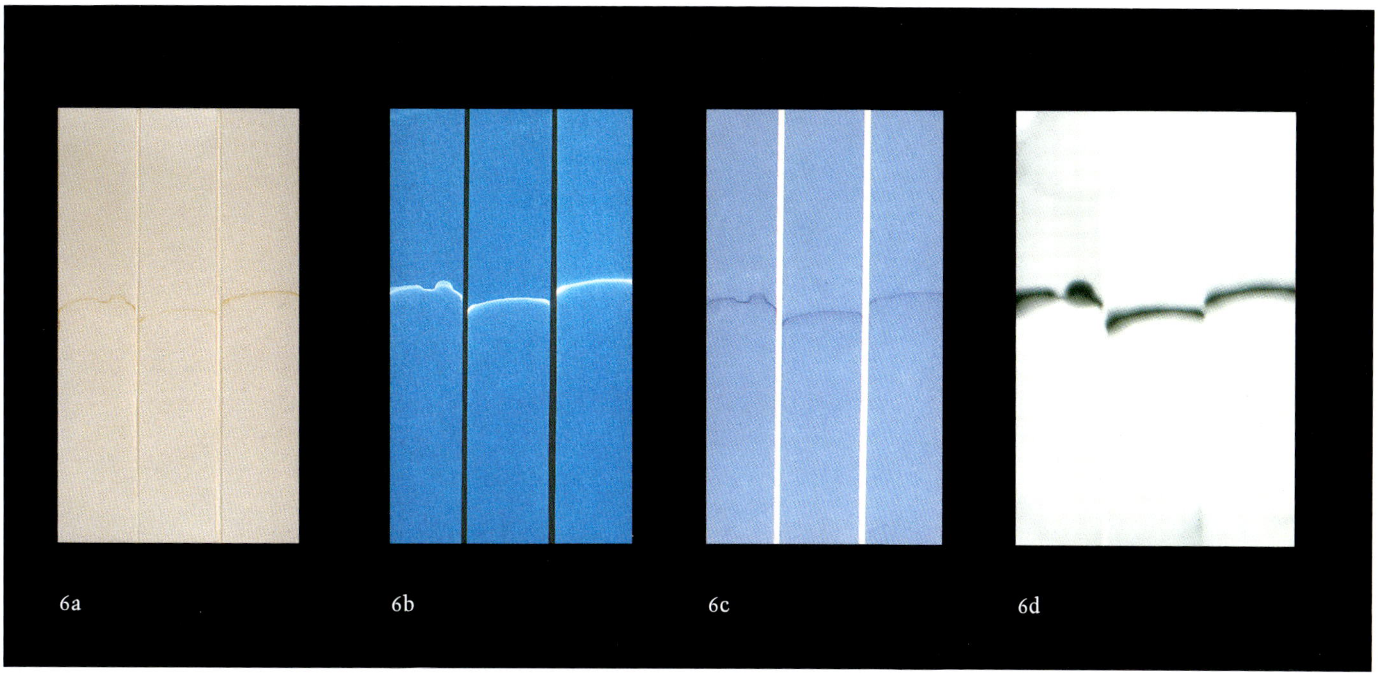

6a 6b 6c 6d

extract small amounts of carbohydrate "impurities" from pure cellulose. This product may be obtained by simple extraction in water at 25°C. After air-drying, the test can be repeated up to fifty times, yielding a small but constant amount of sample after the second or third extraction.[30] In comparison, the tideline experiment produces the same result each time it is repeated on a sheet; the rate of formation of the extractable brown substance does not decrease with repetition.[31]

The ability to make continuous "extractions" could mean that water-soluble degradation products are formed every time the paper dries. The results of the experiments with washed and stored samples demonstrate that water-soluble degradation products are formed even when the paper is completely dry.[32]

Horizontal Sample Experiments

Clearly, the tideline phenomenon has a bearing on partial wetting treatments in paper conservation. However, the test setup as described above hardly compares to real treatments conducted in paper conservation. Therefore, additional tests were designed with two objectives in mind. The first concern was the potential problems associated with the creation of tidelines during local treatment. It has been suspected for some time that fluorescent tidelines that can be seen only under ultraviolet light will discolor in time. The validity of this concern was tested by exposing tideline-containing samples to different accelerated aging techniques (fig. 7).

The second objective was to determine what combination of paper, solvent, and application had the most severe impact on the wet-dry boundary. In choosing materials and applications to meet both objectives, an effort was made to approximate real paper conservation methods so that results from the tests could be of practical value in paper conservation. Thus five papers were treated with five liquids in two methods of application.

Test papers:
For recording purposes, the number or letter in brackets refers to the marks on the samples shown in fig. 8a–b.

COLUMN

A. Whatman 1 filter paper, unwashed [1]
B. Whatman 1 filter paper, washed [5]
C. Kalamazoo, contemporary handmade cotton-linen fiber paper [2]
D. Eighteenth-century handmade rag paper [3]
E. Japanese Nao 10/85, contemporary handmade *kozo* fiber paper [4]

Liquids:

ROW

1. Deionized water [I]
2. Ethanol [II]
3. Aqueous ammonium hydroxide, pH 9 [IV]
4. Aqueous hydrogen peroxide, 2 percent [III]
5. Aqueous sodium borohydride, 0.5 percent [V]

Liquid applications:

1. Eyedropper, 5 drops of liquid [A]
2. Suction disk, 20 drops of liquid [B] (not shown)

Immediately after application of a drop of water on their surface, many naturally aged papers form a yellow-brown tideline through migration of water-soluble compounds to the outer edge. These types of paper are not very useful for determining the relationship between fluorescent and colored products in the tideline. Whatman 1 filter paper was used for the test because the tideline created in this paper is not immediately visible in daylight. In addition, the purity of this paper limits the variables that need to be considered to explain the formation of a tideline. The results of the

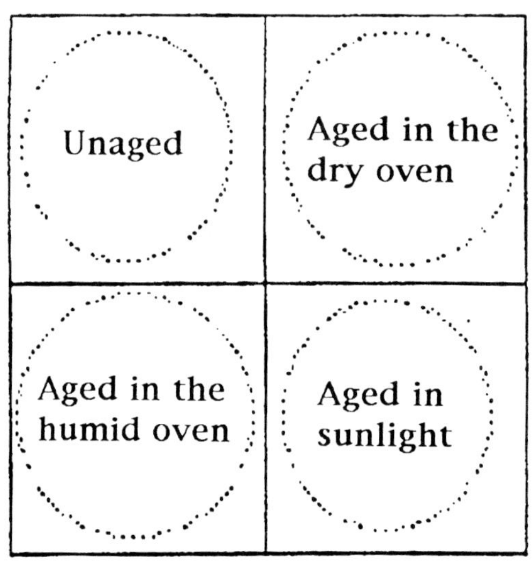

7. Schematic drawing showing how each paper sample was divided and the treatment each part received

tests on Whatman 1 filter paper could easily be compared with other experiments in which the same paper was used. Although its purity suits the requirements for this study, it should be noted that most objects treated in paper conservation will react differently due to their different fiber content and the presence of impurities or additives such as sizing and fillers. Therefore, three other paper samples were included in this test series. Although the observations made on these three samples relate more closely to actual objects, the causes of tideline formation were more difficult to isolate.

The liquids selected were those most frequently used in paper conservation. Water and ethanol were obvious choices. In addition, ammonium hydroxide (pH 9) and an oxidizing and a reducing bleaching agent, respectively hydrogen peroxide (2%) and sodium borohydride (0.5%), were chosen to represent liquids used for more drastic local treatment of stains. The paper samples (4 x 4 in.) were divided into four quarters (see fig. 7). Duplicate setups of the test papers were made. In the first series, each quarter of the paper sample was treated with five drops of liquid using an eyedropper. The samples were air-dried on a fiberglass screen. In the second series, twenty drops of liquid were applied by eyedropper on each paper sample quarter with the sample placed on blotting paper[33] on a suction disk.[34]

Depending on the liquid and kind of paper used, diameters of the wetted areas ranged from 0.25 to 1.5 inches. After drying, each sample was cut into four quarters. To test for discoloration of the tideline area, three of the four quarters were aged for twenty-eight days under the following conditions: one quarter was placed in a dry oven at 80°C, another in a humid oven at 80°C and 50 percent relative humidity, and another in sunlight behind window glass facing south during the month of September. The fourth quarter was reserved as an unaged control.

Results

Figure 8a shows under normal light the twenty-five samples treated with five drops of liquid. The samples are laid out in a grid pattern in which each vertical column represents the different papers, divided into quarters, that were tested; each horizontal row shows five different liquids that were used. Figure 8b shows the same twenty-five samples viewed under ultraviolet light.

Fluorescence before Aging

An increased fluorescence at the tidelines occurs in virtually every sample treated with an aqueous solution (see rows 1, 3, 4, and 5) or in which ethanol (row 2) had been applied with an eyedropper (note the unaged quarters at the top left of every sample in fig. 8). The only exception to the increased fluorescence is the application of ethanol, applied on the eighteenth-century rag paper (column D) and the Japanese *kozo* paper samples (column E), where no tidelines were apparent. The most intensely fluorescent tidelines developed when the hydrogen peroxide solution (row 4) was used. In the unwashed and washed Whatman 1 filter paper samples (columns A and B), this solution created double or broad rings around and within the treated area.

While the tideline region contained the most fluorescent material, the region outside the treated circle generally contained more fluorescent material right after application of the liquid than the region inside the treated circle. When the suction disk was used, the removal of fluorescent material within the treatment area was evident. This second observation suggests that migration and concentration of water-soluble fluorescent material at the wet-dry boundary is a crucial factor in the creation of increased fluorescence in the tideline region. This assumption was supported by the fact that washed and then air-dried Whatman 1 filter paper samples had lost a small amount of fluorescence over the entire surface and had dimmer tidelines than the unwashed samples.

Using the suction disk resulted in far fewer test paper samples in which fluorescent tidelines were visible under ultraviolet light even though more liquid had been applied (20 drops) compared to the eyedropper (5 drops). The Kalamazoo and Japanese paper samples rarely contained tidelines after they were treated on the suction disk.

Fluorescence after Aging

After aging three of the four quarters of the samples, a clear difference in fluorescence of the surface of the paper could be observed. The

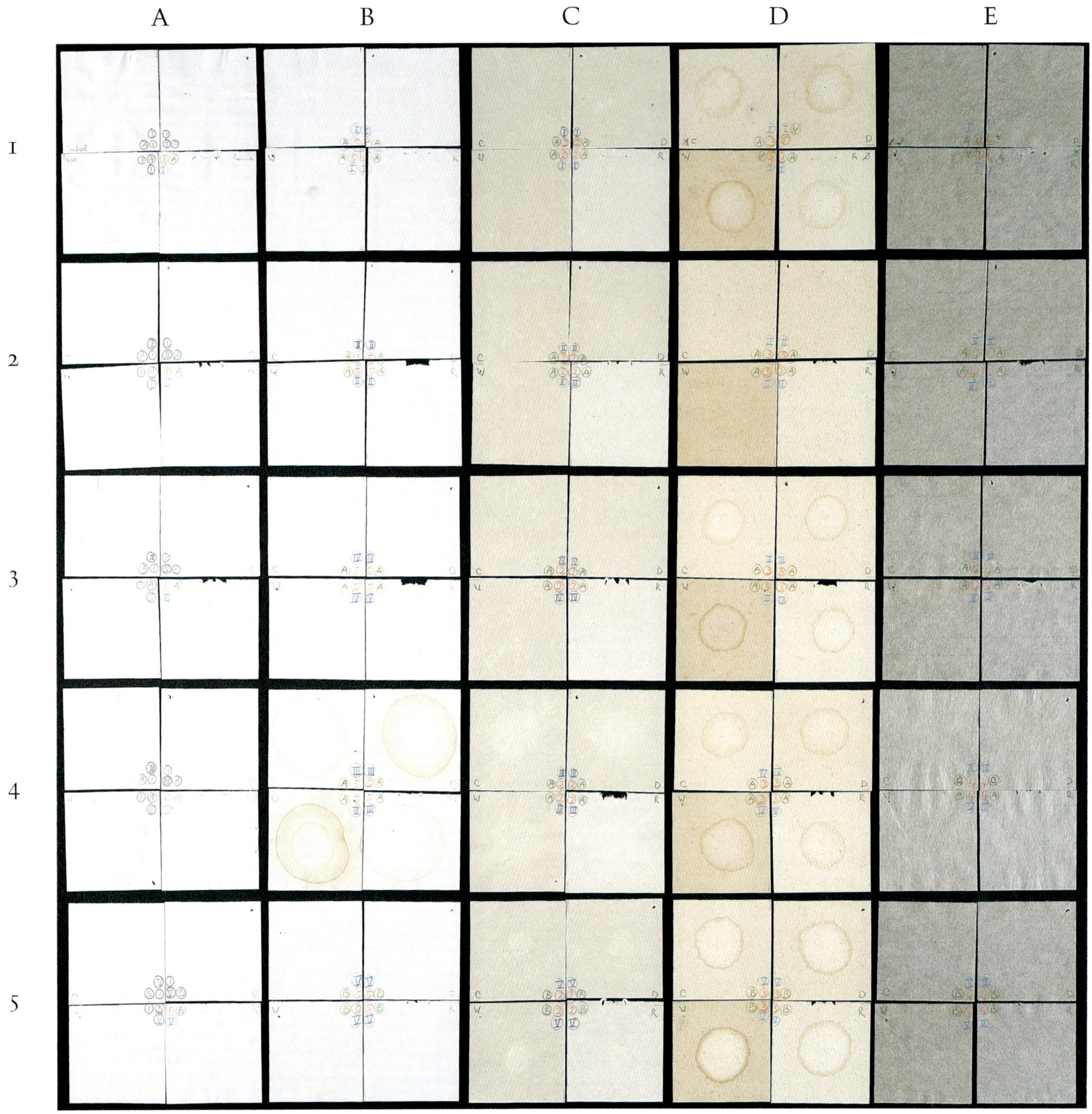

8a. Test papers (columns A–E) viewed in daylight following application of different liquids (rows 1–5) and aging conditions

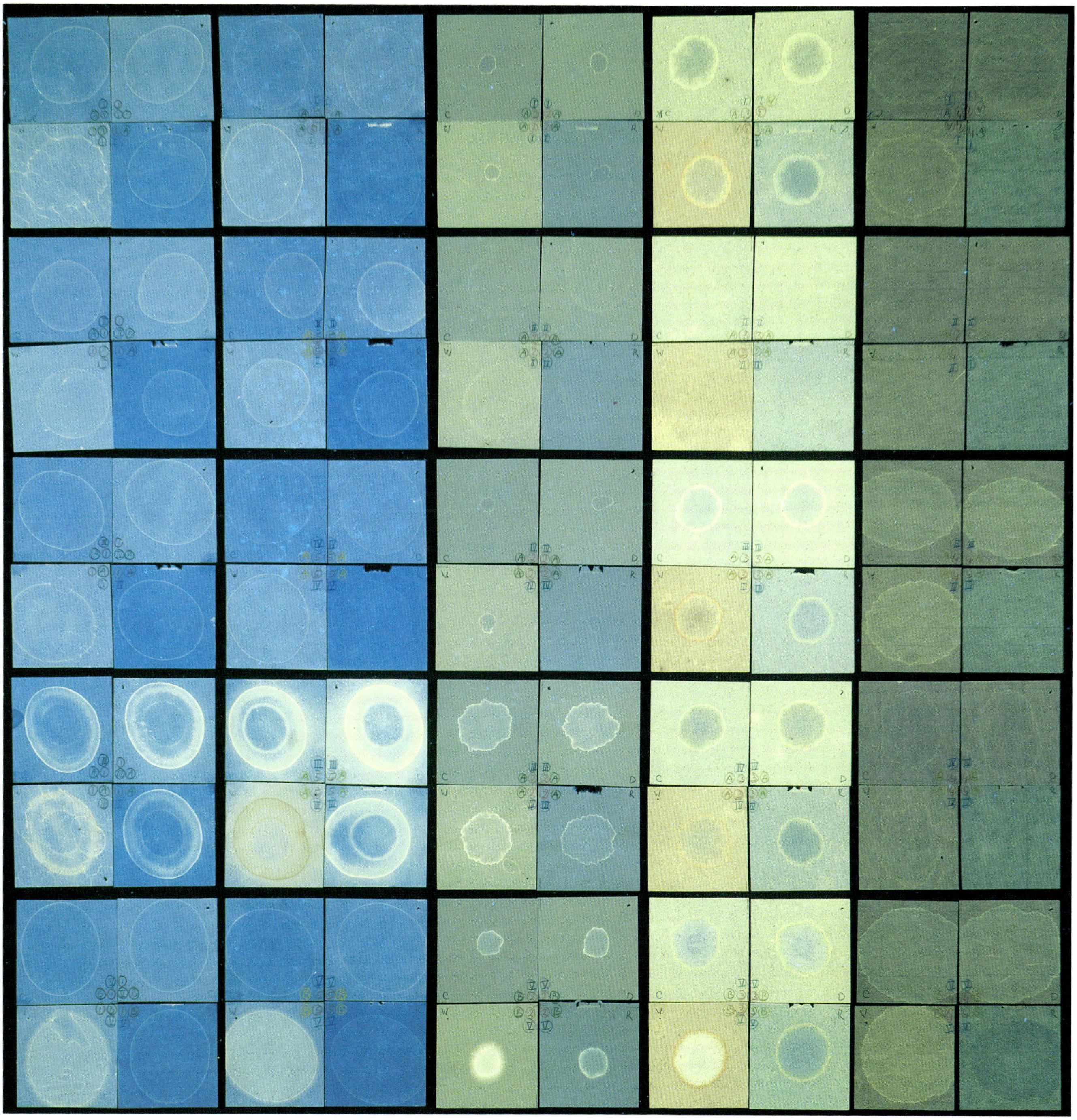

8b. Test papers shown in fig. 8a viewed under ultraviolet light

test sample quarter that was aged in a humid oven, in general, contained the most fluorescent material on its entire surface. The quarter that was aged in a dry oven contained less fluorescent material than the humid-aged quarter but more than the control quarter. The light-aged quarters, in general, contained the least amount of fluorescent material. Not only the fluorescence intensity of the paper surface changed after aging but also the fluorescence color. The humid-aged samples tended to show a yellow fluorescence, while the dry-aged samples fluoresced light yellow. The yellow fluorescence was very visible on the Kalamazoo and eighteenth-century rag paper samples and to a lesser degree in the unwashed and washed Whatman I filter papers. The light-aged quarters, in general, not only lost a large portion of their fluorescence but also appeared more blue than the other three quarters. This result pertained for all the different papers and particularly for the eighteenth-century rag paper.

The shifts in intensity and color of fluorescence among the four quarters were even larger in the tideline region and ranged from white-blue fluorescence to dark-brown nonfluorescence. Like the overall surface, the tideline in the humid-aged quarter contained the most fluorescent material, followed by the dry-aged quarter and the control quarter. Also, like the overall surface, the fluorescence of the tideline decreased when exposed to sunlight for twenty-eight days. The tideline in the unaged quarters was generally blue under ultraviolet light. Aging the samples in a dry oven resulted in only a slight color shift of the tideline fluorescence to yellow, but aging the sample in the humid oven produced a strong color shift to yellow. In a small number of samples the fluorescence had almost completely disappeared and was replaced by a brown coloration.

The yellowing of cellulose caused by artificial aging shows parallels with artificial aging and yellowing of other organic materials. Organic molecules with sufficiently long systems of alternating single and double bonds (conjugated system) can absorb ultraviolet light and may fluoresce. Researchers have suggested that as degradation progresses, these compounds can produce more double bonds. These longer conjugated systems can have the capability of absorbing light from the violet end of the visible spectrum, thus appearing as yellow material.[35] Studies on the fluorescence of varnish layers produced results comparable to the results obtained in this research. E. René de la Rie showed that dried linseed oil film produces an increase in fluorescence upon continued storage in a dark environment. Combined with an increase in fluorescence, the color of the fluorescence shifted from bluish to yellowish.[36] In a more recent study de la Rie also found that light with a wavelength greater than 400 nanometers can break down discolored products in films of dammar.[37]

Discoloration before Aging

Before aging, tidelines were only visible in daylight on the eighteenth-century rag paper treated with aqueous solutions using an eyedropper. The liquid transported a large amount of the discolored material to the outside edges, resulting in a lighter-color treated area and a dark tideline. The only other combination of paper and liquid that produced a visible tideline before aging was washed Whatman I filter paper that contained hydrogen peroxide.[38]

Discoloration after Aging

Aging the samples in the humid oven had not only changed the color of the fluorescence of the tideline region, but in daylight faint traces of yellow discoloration began to appear in all unwashed and washed Whatman I filter paper samples.[39] In cases of severe degradation, for example in the Whatman filter paper sample treated with hydrogen peroxide and aged in the humid oven, the tideline was nonfluorescent under ultraviolet light and intense brown in daylight. In general, the blue fluorescent tidelines were not visible in daylight, and yellow fluorescent tidelines showed traces of discoloration in daylight.

This sequence in fluorescence and color shifts in the tideline region indicates an increase in degradation in the series control quarter, dry-oven–aged quarter, and humid-oven–aged quarter. As degradation proceeds, the fluorescence intensity increases, the fluorescence color changes from blue to yellow, and visible discoloration begins to appear. Eventually, fluorescence disappears, and the tideline appears dark in ultraviolet light.

This sequence of fluorescence and discoloration applies not only to the tideline region but also to fluorescence in paper in general. After the eighteenth-century rag paper samples aged in the humid oven, former randomly present fluorescent patches became visibly discolored. Other papers that contained fluorescent spots but had no discoloration visible in daylight were also aged in the dry oven. After aging, the fluorescence had disappeared and the spots had discolored to become what are generally characterized as "foxing spots." Consequently, the formation of a visible tideline does not appear to be a singular phenomenon. Fluorescence in paper may be an indication of early stages of degradation.

The effect of light aging was the opposite of aging in both the dry and humid ovens. Discoloration as well as fluorescence either decreased or disappeared completely after exposure to the sun for twenty-eight days. The yellow fluorescent compounds were broken down faster than the blue fluorescent compounds, producing a color shift of the fluorescence from yellow to blue. It should be emphasized, however, that sunlight does not reverse the process of discoloration. The compounds responsible for the discoloration of the samples in these experiments are most likely products of paper degradation processes, which can include chain and ring scission and cross linking in the cellulose molecules.[40] Apparently, exposing degradation products to sunlight breaks them down to molecules incapable of absorbing light, which are, therefore, colorless. The paper will not return to its original state following exposure to sunlight.

Influence of Paper, Liquid, and Method of Application

The darkest tidelines appeared after aging in the humid oven on Whatman 1 filter paper treated with hydrogen peroxide and on eighteenth-century rag paper treated with any aqueous solution. The tidelines created using deionized water on washed Whatman 1 filter paper were not as dark as the tidelines on the unwashed Whatman 1 filter paper; however, they were still visible after aging in the humid oven. In washed and unwashed Whatman 1 filter paper samples, tidelines created by an ethanol application were very faint but still visible after aging. Ammonium hydroxide, in general, reacted in the same manner as water on the different paper samples. Sodium borohydride showed a significant increase in fluorescent material; however, none of the samples treated with this solution had visible discoloration after aging.

Methylene Blue Absorption

From observations from previous experiments by this author, it was expected that fluorescent tidelines would occur in tests described above. It seemed unlikely, however, that the same amount of degradation would take place as in the vertical suspended strips because of the much smaller amount of liquid evaporating from the interface. A separate series of Whatman filter paper samples was treated with five drops of deionized water and tested for methylene blue absorption. Samples treated with deionized water, deionized water and aged in a dry oven, and deionized water and rinsed for three hours were all subjected to the methylene blue absorption test.

The separate series of Whatman 1 filter paper samples were treated aged and unaged by immersion for two minutes in a 0.1 percent methylene blue solution. The samples were then rinsed for two hours. Samples aged in the dry oven showed fluorescence under ultraviolet light and discoloration in daylight. They also clearly showed increased absorption of the blue dye in the tideline region. The tidelines in the unaged samples showed enhanced fluorescence but no visible discoloration. Although far less methylene blue was absorbed by the unaged samples, the tideline was still very visible because of an increased absorption of methylene blue. The unaged samples that were rinsed in deionized water after application of the five drops of HPLC-grade water gave no indication of enhanced fluorescence under ultraviolet light. However, when the immersion was closely monitored and excessive dye thoroughly rinsed out, traces of enhanced methylene blue absorption remained visible in the tideline region. It appears, therefore, that some change takes place at the wet-dry boundary in Whatman 1 filter paper as soon as five drops of HPLC-grade water evaporate from its surface, although no fluorescence and discoloration occur.

Conclusion

Several tests established that a cellulose degradation process takes place in the tideline region. Increased methylene blue absorption and formation of water-soluble fluorescent and colored compounds occur at the wet-dry boundary of both the vertical suspended samples and the horizontal samples. A combination of the migration of paper degradation products, among which are peroxides and the swollen fibers at the wet-dry boundary that provide easy access of oxygen to the cellulose, could account for the degradation.

The results of the experiment in which the samples were washed and stored indicate that some cellulose degradation occurs continuously. Russell-effect photography showed the generation of new peroxides at every wet-dry boundary.

In Whatman 1 filter paper, fluorescent tidelines created with five drops of water have a clear potential to become visibly discolored in daylight during aging. The formation of fluorescent and colored products in paper is not a unique process and can be related to comparable degradation processes in other organic materials.

Although the results of the tests conducted using Whatman 1 filter paper indicate a high potential for the fluorescent tideline to become discolored, objects normally treated by the paper conservator rarely match the structure and texture of filter paper. Many factors, including fillers and sizing, might inhibit tideline formation or its transition from fluorescence to discoloration. Kalamazoo paper, for example, did not show the dramatic changes that appeared in Whatman 1 filter paper. Gelatin sizing, fungus, and optical brighteners can all cause fluorescence in paper; not all the fluorescence observed under ultraviolet light is caused by paper degradation products. Consequently, not all fluorescent regions have a potential for discoloration.

During the course of this research it was observed that the fluorescence in dry samples was transferred to papers that were in direct contact with the original sample. This transfer occurred only when the sample and paper were kept in dark storage for a longer period of time (1 to 2 years) and were not disturbed.

Practical Implications

The most significant consideration resulting from this research is the importance of examining objects under ultraviolet light before and after treatment. Close monitoring of the treated area under ultraviolet light should be employed when hydrogen peroxide or sodium borohydride is used during conservation treatment.

Using a suction disk during the application of liquids to paper reduces the potential for tideline formation. Although the amount of liquid applied in tests using the suction disk was four times higher than when using the eyedropper alone, the tidelines around the treated area were smaller in diameter and contained less fluorescent material. Allowing the area to remain wet for shorter periods of time, working at a steady pace, and changing the blotting paper often between the object and the suction disk greatly improve the chances of avoiding tidelines.

Working in a 100 percent relative humidity environment also greatly reduces the possibility of creating tidelines during local treatment. The object should be in a humid environment long enough to acclimate and take up moisture from the surrounding atmosphere. When the paper is saturated with water, the formation of a wet-dry boundary during local treatment will not occur. Humidifying the object in a cold-humidification chamber and treating the object on wet blotting paper covered with Gore-Tex proved to be successful in avoiding tidelines.

Tidelines created by organic solvents are generally of a different nature. For example, a tideline that is created using xylene is removable by immersing the paper in the same solvent. However, contrary to what occurs when an aqueous solution is used, the former tideline region will not appear through an enhanced methylene blue absorption. This result may indicate that degradation of the solvent rather than the paper is taking place at the wet-dry boundary. The risk of creating tidelines when using organic solvents for local treatment can be reduced when the treatment takes place in an environment saturated with the solvent.

The areas inside and outside the tideline in all test paper samples were bleached when exposed to sunlight. Both the overall surface of the paper sample and the tideline also lost a large amount of their fluorescence. As a curative measure, exposure to sunlight or light bleaching would appear to be an effective method to remove tidelines and general discoloration. It also causes a decrease in fluorescent material, which could discolor during aging. However, sunlight works as a bleaching agent and attacks cellulose molecules. The sunlight-aged samples also showed that this procedure rarely removes all fluorescent material.

Data from the textile industry also indicate that a tideline in cotton cellulose can be removed immediately after formation. By comparison, if a paper object is dried, the removal of the discoloration will be more difficult. Drying at higher temperatures often makes the discoloration impossible to remove.[41]

NOTES

I wish to thank the Mellon Foundation for funding my position in the conservation division of the National Gallery of Art. I would also like to thank the following National Gallery staff members for their assistance: Pia DeSantis Pell, former senior paper conservator, whose research on enzymes and poultices prompted me to start this project; Shelley Fletcher, head of paper conservation; Judy Walsh, senior paper conservator, who gave me many helpful suggestions and added greatly to the final format of the publication; E. René de la Rie, head of the scientific research department, who gave me many suggestions and helpful insights throughout the course of this research; Suzanne Quillen Lomax, organic chemist, who carried out the peroxide analysis; and Constance McCabe, senior photograph conservator, who gave me advice and assistance in obtaining the film used for the Russell-effect process. Philip Charles, of the photograph and visual imaging division, gave indispensable help with photographing the test papers. I also wish to thank Vincent Daniels, head of the conservation department of the British Museum, London, for his time and practical advice concerning Russell-effect photography.

The epigraph is from D. Heinz Corte, *Handbook of Paper Science* (Amsterdam, 1980), 80. Corte's conclusion is preceded by the following analysis:

Even if the mechanism or mechanisms of "hornification" were known it would still leave the problem of what to do about them. De Ruvo and his coworkers found that a partial acetylation (3–4 percent by weight) improves the recyclability of pulp, presumably because the acetate groups prevent hydroxyl groups in their neighborhood from recombination. This is perhaps not a realistic proposition for making recyclable paper but the ultimate aim of research in this field must be to find ways and means to reduce the damage done to the fibers as much as possible, that is, to make paper with recycling in mind. This, of course, clashes with the efforts of generations of paper makers to increase the utility of many grades of paper by increasing their resistance to water.

1. Jane K. Hutchins, "Water-Stained Cellulosics: A Literature Review," *Journal of the American Institute for Conservation* 22 (1983), 57–61; Frank J. Ligterink, Henk J. Porck, and Wim J. T. Smit, "Foxing Stains and Discoloration of Leaf Margins and Paper Surrounding Printing Ink," *Paper Conservator* 15 (1991), 48.

2. W. H. Bone, "Evaporation of Water from Cellulose," *Journal of the Society of Dyers and Colourists* 50 (1934), 307–309.

3. Dianne van der Reyden, "Recent Scientific Research in Paper Conservation," *Journal of the American Institute for Conservation* 31 (Spring 1992), 129.

4. W. H. Bone and H. A. Turner, "Some Effects of the Evaporation of Water from Cellulose," *Journal of the Society of Dyers and Colourists* 66 (1950), 315–327; Herman Bogaty, Kenneth S. Campbell, and William D. Appel, "Some Observations on the Evaporation of Water from Cellulose," *Textile Research Journal* 22 (February 1952), 75–81; G. W. Madaras and H. A. Turner, "Further Observation on the Effects of Evaporating Water from Cotton Cellulose," *Journal of the Society of Dyers and Colourists* 69 (October 1953), 371–377; Robert Schaffer, William D. Appel, and Florence H. Forziati, "Reactions at Wet-Dry Interfaces on Fibrous Materials," *Journal of Research at the National Bureau of Standards* 54 (February 1955), 103–106; M. R. Fox, "Wet and Dry Boundaries on Cellulosic Textiles and Their Influence on Dyes and Dyeing," parts 1 and 2, *Journal of the Society of Dyers and Colourists* 81 (1965), 7–11, 46–51.

5. T. P. Nevell and S. H. Zeronian, *Cellulose Chemistry and Its Applications* (New York 1985), 247. Methylene blue is an indicator used for determining the amount of carboxyl groups in cellulose. It has been shown in experiments that with the use of buffers the methylene blue absorption and carboxyl content are equal at a final pH of about 7.8. The methylene blue concentration can be measured spectrophotometrically at 625 nm. In addition to quantitative use, the dye is an excellent visual tool for indicating the presence of carboxyl groups in a cotton cellulose sample.

6. The sample can be immersed for 10 minutes in a 0.1 percent solution of methylene blue. After rinsing the paper for 1 hour, the sample will clearly show increased absorption of methylene blue in the tideline region.

7. Bone and Turner 1950, 316.

8. Bone and Turner 1950, 319. Harrison's solution is a mixture of sodium thiosulfate, silver nitrate, and sodium hydroxide. Bone's sample was immersed for 3 minutes in the solution, which had been heated to 90°C. The sample was rinsed in sodium thiosulfate and then in water. If the sample was sufficiently purified beforehand, it showed a faint pink appearance. The tideline region, however, was visible as a strong black-brown coloration. Like methylene blue absorption, the coloration using Harrison's solution was observed in both instances—with or without the

yellow-brown discoloration of the tideline having been washed out before conducting the test.

9. The cuprammonium fluidity is a method for determining the viscosity of cellulose in a metal complex solution. It has been found that the fluidity of the solution linearly increases with a percentage fall in fiber strength.

10. Bone 1934, 308.

11. Madaras and Turner 1953, 371–377, conducted the tideline test in a nitrogen atmosphere (partial oxygen pressure was reported to be 4×10^{-7} of the total pressure) and in vacuo. Even in a nitrogen atmosphere, when traces of oxygen have been rigorously removed, a tideline still appears in the sample. They concluded that if oxygen is indeed necessary for the formation of a brown line, the required amounts are minute. The authors managed to carry out the experiment without the formation of a brown line only in vacuo. However, staining of the strip in the boundary area still showed an increased methylene blue absorption. Although the replication of these tests appeared to be poor, one experiment seemed valuable. As soon as oxygen was released into the system, the tideline area turned yellow-brown.

12. Schaffer, Appel, and Forziati 1955, 103–106. In addition to cotton, nylon and cellulose acetate are also capable of acquiring tidelines when water is used in the test setup. Replacing water with organic solvents such as benzene, xylene, chloroform, and n-pentane resulted in clear demarcations on cotton cloth within 16 hours. The brown, fluorescent products formed in cotton when benzene, xylene, and n-pentane were used were not soluble in water but only in the organic solvent. When, for instance, the brown products formed by using xylene are extracted with xylene, the tideline area is not only nonfluorescent but shows no increase in methylene blue absorption. The authors suggested that in these cases it was the solvent itself that had reacted in the wet-dry boundary and not the cellulose substrate. In other words, a fluorescent tideline, visible after local treatment with organic solvents, could be the result of oxidation of the solvent at the wet-dry boundary. It remains unclear whether the solvent oxidizes only at this point or over the whole wetted area and is concentrated at the wet-dry boundary.

13. Fox 1965, 9. Commercially purified cotton, mercerized cotton, linen, secondary cellulose acetate, and cellulose triacetate were tested with water for tideline formation. Brown, fluorescent tidelines formed in the cotton and linen samples. A tideline that formed in the diacetate sample was faint brown and fluorescent; however, nothing was detected in the cellulose triacetate. Fox concluded that both primary and secondary hydroxyl groups in cellulose take part in the degradation and discoloration of the substrate.

14. Fox 1965, 7.

15. Bogaty, Campbell, and Appel 1952, 78.

16. Fox 1965, 7.

17. Bogaty, Campbell, and Appel 1952, 79.

18. Fox 1965, 7.

19. Fox 1965, 7.

20. Fox 1965, 10.

21. Bone and Turner 1950, 326.

22. Suzanne Quillen Lomax, "Analysis of Peroxides," 5 May 1993, and "Analysis of Tidelines," 3 September 1993, National Gallery of Art, Washington, scientific research department.

23. Vincent Daniels, "The Russell Effect," *Studies in Conservation* 29 (1984), 57–62; Daniels, *Monitoring the Autoxidation of Paper Using Photographic Materials*, Advances in Chemistry Series, no. 212 (Washington, 1986), 318–327; Daniels, "Oxidative Damage and the Preservation of Organic Artifacts," *Free Radical Research Communications* 5 (1989), 213–220.

24. Film that is suitable for Russell-effect photography is difficult to obtain. Generally most types of film contain stabilizers that protect the film from "fogging," a term used for what is the result of oxidation of the silver halide, resulting in faint gray patches on the film. This process is thought to be similar to the Russell-effect. Silver gelatin film containing these stabilizers is therefore unsuitable for our purposes.

25. Treatment with ammonium hydroxide increases the sensitivity of photographic film by increasing the silver halide grain size. Several concentrations of ammonium hydroxide have been suggested in the literature. At the National Gallery of Art, a solution of 0.05 M worked satisfactorily. The film was immersed in ammonium hydroxide for 4 minutes. The film was then rinsed for 2 minutes in distilled water and an additional 2 minutes in ethanol. This process, as well the air-drying of the film, took place in the dark. After the film had dried, it was brought into direct contact with the sample and stored in a lightproof box for 24 hours. The exposed film was developed in Kodak developer D76 for 4 minutes and fixed with Kodak RapidFix for 6 minutes.

26. William J. Russell, "On the Action Exerted by Certain Metals and Other Substances on a Photographic Plate," *Proceedings of the Royal Society* 61 (1897), 424–433; Russell, "On Hydrogen Peroxide as the Active Agent in Producing Pictures on a Photographic Plate in the Dark," *Proceedings of the Royal Society of London* 64 (1899), 409–419; Russell, "On the Action of Wood on a Photographic Plate in the Dark," *Proceedings of the Royal Society of London* 74 (1905), 131–134.

27. Guilleminot EG 18, made in France. This film is available through Chicago Albumen Works, P.O. Box 379, Front Street, Housatonic, Mass. 01236.

28. Vincent Daniels, British Museum, London, personal communication.

29. In the following experiments it was hoped that water-soluble antioxidants would inhibit cellulose degradation at the wet-dry boundary. L-ascorbic acid and ascorbic acid calcium salt are both well-known antioxidants; some interesting results using these

compounds were reported in S. Allen Grimsley, James C. Robinson, and Mark A. Schroeder, "Enhanced Dithionite Bleaching," U.S. pat. 4,871,423, 3 October 1989. Color reversion after bleaching the pulp with hydrogen peroxide decreased when the bleaching agent was used in combination with ascorbic acid. A water-soluble phenolic antioxidant was provided by Ciba-Geigy. The three antioxidants were tested in the test setup described above using Whatman 1 filter paper. The filter paper was exposed to the test for 24 hours using different concentrations of antioxidants. After 24 hours both the L-ascorbic acid and the ascorbic acid calcium salt samples showed large areas of severe discoloration. The tideline region was deep brown and not contained within a narrow band. The whole wetted area was yellow-brown. The phenolic antioxidant did not show this dramatic color change, but neither did it inhibit the formation of discoloration at the tideline region.

30. J. Strachan, "Solubility of Cellulose in Water," *Nature* 141 (19 February 1938), 332–333; G. W. Huffman, P. A. Rebers, D. R. Spriesterbach, and F. Smith, "Nature of a Hemicellulose Extracted from Cellulose with Water," *Nature* 175 (4 June 1955), 990–991.

31. Bogaty, Campbell, and Appel 1952, 76.

32. Corte 1980, 72. A phenomenon that could be relevant to these observations is called "hornification" in the paper industry. It is a vague term used for the effect overdrying has on paper (drying at high temperatures). The term is also used to indicate the surface of paper: a "horn-layer" of oxidized cellulose.

33. James River Verigood blotting paper.

34. The suction disk system contained a stainless steel photoetched disk, 0.7 mm (700 micron), and a Gast oilless rotary vane vacuum pump, model 0323.

35. Vincent Daniels, "The Discoloration of Paper on Aging," *Paper Conservator* 12 (1988), 93–99.

36. E. René de la Rie,"Fluorescence of Paint and Varnish Layers," part 2, *Studies in Conservation* 27 (1982), 66–69.

37. E. René de la Rie, "Photochemical and Thermal Degradation of Films of Dammar Resin," *Studies in Conservation* 33 (1988), 53–70.

38. The high concentration of peroxides seems to have initiated cellulose degradation reactions in which colored products were formed.

39. This yellow discoloration was not only faint, but the yellow line was also very narrow and therefore detectable only with the eye and not with the use of a colorimeter, densitometer, or spectrometer. For the same reason, the line could be reproduced photographically only using black-and-white film with blue filters. Aging a Whatman filter paper sample after application of 5 drops of deionized water in a dry oven at 175°C causes the formation of an intense brown tideline after 20 minutes.

40. William K. Wilson and E. J. Parks, "An Analysis of the Aging of Paper," *Restaurator* 3 (1979), 37–61.

41. Fox 1965, 7.

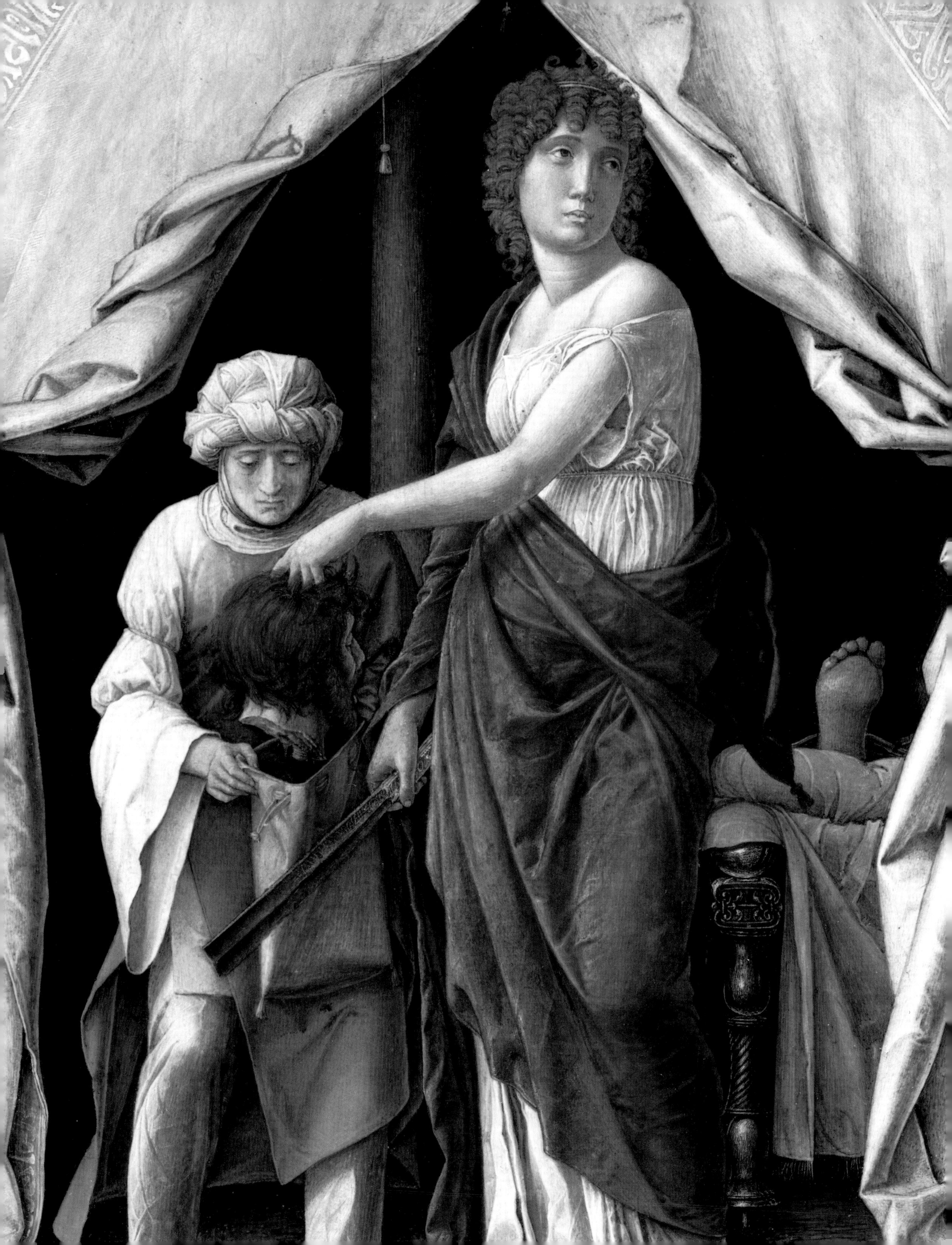

SUSANA M. HALPINE

An Investigation of Artists' Materials Using Amino Acid Analysis: Introduction of the One-Hour Extraction Method

Understanding the nature of the materials artists have used is essential for conservation treatments. Identification of binders such as tempera, distemper, and casein, applied in layers or as mixtures, may also reveal the original intent of the artist, give insight into the artist's techniques, and help explain why painted surfaces are affected differently over time. While such proteinaceous binders cannot be characterized with any degree of certainty using existing analytical methods such as fatty acid analysis or cross-sectional staining, the sensitive amino acid analysis technique described in this paper can identify very small amounts of such material.

For example, at the outset of this project, both animal glue (distemper) and egg tempera were identified as paint binders on a fifteenth-century Italian panel painting,[1] indicating the artist's preferential use of binders to create specific optical effects for various colors. Glue was recommended as a binder for blue and some green pigments by Cennino Cennini in his fifteenth-century handbook, *Il Libro dell'Arte*,[2] but this mixed media technique has never been confirmed for lack of a suitable analytical method.[3] Therefore amino acid analysis can also help confirm or refute artistic techniques described in the literature. Furthermore, animal glue is hygroscopic, and the finding of the glue binder alongside the egg tempera helped explain why the paint lifted in certain blue areas of the panel painting but remained attached in other areas.

The primary goal of my research was to exploit high-performance liquid chromatography (HPLC)[4] in general and amino acid analysis in particular for possible applications in art conservation (see appendix). HPLC is a highly flexible analytical technique ideally suited for a range of artists' materials, from lake pigments, oils, and resins to carbohydrates and proteins. I chose to analyze proteinaceous artists' materials because there is a lack of reliable methods in the conservation field[5] and because of my background in protein chemistry. An amino acid analysis system, capable of identifying extremely small amounts of material, was successfully adapted to the requirements of conservators. This system is now routinely used to address technical questions regarding the nature of materials found on art objects undergoing treatment at the National Gallery of Art.

In this essay the search for improved analytical methods will be reviewed and selected results from technical investigations of paintings undergoing conservation treatment will be discussed. Certain techniques emerged that will prove helpful to conservation scientists using amino acid analysis, while other techniques can be further explored in future research. The results confirm some techniques previously described in the literature while refuting others; at times new techniques are suggested. The analytical results from works by Agnolo Gaddi (active 1369–1396), Andrea Mantegna (1431–1506), Edward

Coley Burne-Jones (1833–1898), Paul Gauguin (1848–1903), and Henri de Toulouse-Lautrec (1864–1901) will also be discussed.

Aside from the scientific data, this research suggested two observations that will not be addressed in depth but are worth mentioning. The first was the prevalence of the use of animal glue, in its many forms and applications, throughout the history of art. Analytical results often confirmed documentation of the use of glue and also indicated the presence of glue in undocumented applications. Glue has been identified as the paint binder on ancient Egyptian wooden objects,[6] on a Florentine Renaissance festival canvas backdrop,[7] and on an eighteenth-century New Mexican *retablo*.[8] Glue was used for specific colors on early Renaissance Italian "tempera" panels;[9] and in the twentieth century Mark Rothko was known to have used rabbit skin glue to give a "subtle-sparkle" to his chapel murals.[10] Glue was also extensively used in carpentry; as a size to seal canvas, paper, stone, and wood before painting; mixed with calcium sulfate or calcium carbonate to prepare the gesso ground of paintings and moldings on frames; as a varnish coating; and later, by conservators for consolidating flaking paint, relining a canvas, and inpainting paint losses.[11] Pliny the Elder in the first century, Theophilus in the twelfth century, and Cennini in the fifteenth century described glue recipes for specific applications.[12] In the eighteenth century Henri Louis Duhamel du Monceau recommended glues made from the hides of older animals for strong glue applications such as carpentry ("the older and leaner the animal, the stronger the glue"), while weaker but optically clearer glues from young animal skins were recommended for painting.[13] Parchment as well as sturgeon glues are still sought for their clarity for many applications in art conservation. Although once widely used, animal glues have been replaced by synthetic adhesives in most applications because glues are hygroscopic and can weaken and peel over time.[14] Amino acid analysis, in combination with procedures such as the extraction method described below, should help identify other specific glue applications in works of art.

During my research, it became clear that a number of artists appear to have been innovative in the use of their materials. Despite the assumption that techniques were passed on as workshop traditions from one generation to another, it is not inconceivable that certain techniques were guarded as trade secrets in order to compete for patronage.[15] Amino acid analysis should be helpful in identifying ways in which artists went beyond traditional techniques and materials in their search for creative processes better suited to the effects they wished to achieve.

Many of the technical inquiries described here are still in progress. Confirmation of interpretations based on amino acid analysis results has been and continues to be sought through other scientific techniques as well as corroborated with art-historical references.

Background

Analytical Technique

The amino acid analysis consists of breaking down a proteinaceous sample into its constituent amino acids during exposure to a hydrochloric acid vapor, in an oven at 110°C for twenty-four hours. The individual amino acids are subsequently labeled for identification on the HPLC by derivatization with

Amino Acid Symbols

D=Aspartate	NH=Ammonia	RE=Reagent (PITC)
E=Glutamate	Y=Tyrosine	K=Lysine
OHP=Hydroxyproline	V=Valine	CA=Cysteic acid
S=Serine	M=Methionine	MS=Methionine sulfone
G=Glycine	C=Cystine	PS=Phosphoserine
H=Histidine	I=Isoleucine	PT=Phosphothreonine
R=Arginine	L=Leucine	GAL=Galactosamine
T=Threonine	NL=Norleucine (internal standard)	GLA=Glucosamine
A=Alanine	F=Phenylalanine	W=Tryptophan
P=Proline		?=Unknown

phenylisothiocyanate (PITC) to create phenylthiocarbamyl (PTC)-amino acids. The PTC-amino acids are then separated on the HPLC, identified, and quantified (fig. 1a). The composition of the unknown sample is then compared to the composition of known materials (table 1). Some PITC reagent remains in the sample and elutes at approximately 12 minutes. An internal standard, norleucine, added before hydrolysis to monitor the amount of the sample, elutes at 11.2 minutes.[16] The system used here is based on the Pico-Tag amino acid analysis system developed by Waters Associates.[17] The HPLC instrument is a Hewlett Packard 1090M, equipped with a diode array detector monitoring at 254 nanometers, and a computer for data processing.[18]

There are four major proteinaceous artists' materials: gelatin or animal glue, casein, egg yolk, and egg white (glair). Table 1 compares the amino acid composition of these binders. Although the compositions are generally distinctive, samples frequently contain mixtures of these binders, such as egg tempera over a gesso ground containing animal glue. Therefore specific markers—components unique to the material or present at distinctive levels—were sought to identify each of the proteinaceous materials when amino acid compositions overlapped.

fig. 1a

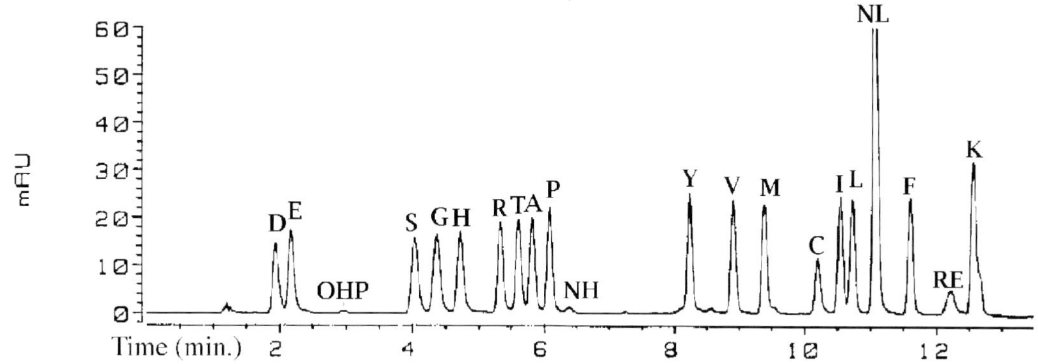

fig. 1b

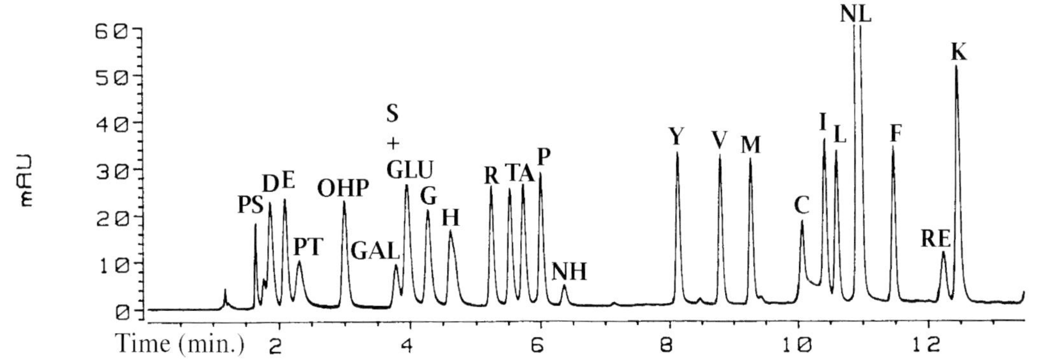

1. Chromatograms of an amino acid reference standard derivatized with phenylisothiocyanate (PITC); 200 picomoles of each phenylthiocarbamyl (PTC)-amino acid, and one nanomole of PTC-norleucine, the internal standard. (a) The regularly monitored proteinaceous PTC-amino acids are, by elution order: aspartate, glutamate, serine, glycine, histidine, arginine, threonine, alanine, proline, phenylthiourea (a product of PITC and ammonia), tyrosine, valine, methionine, isoleucine, leucine, norleucine, phenylalanine, PITC reagent, and lysine. (b) Modified PTC-amino acids such as phosphoserine, phosphothreonine, glucosamine, galactosamine, and hydroxyproline as well as PTC-amino acid tryptophan (usually destroyed during acid hydrolysis) were run together to show chromatographic placement with the proteinaceous PTC-amino acids

Common Proteinaceous Materials

Gelatin and animal glue are both composed of collagen, the main protein in animal structural organs such as skin and bones. Fortunately, the amino acid hydroxyproline is a fairly reliable marker for this ubiquitous artists' material. In the animal kingdom, hydroxyproline is found almost exclusively in collagen[19] (see table 1, columns 1–2, for amino acid composition). Therefore, when hydroxyproline appears in a sample, we can use it as a specific marker to infer that glue is present and thereby identify glue even in complex mixtures. Characterization of animal glue is confirmed by a high (30+) percentage of glycine, a 1:8 ratio of serine-to-glycine, an almost equal alanine-to-proline ratio, and low amounts of essential amino acids such as valine, phenylalanine, and leucine. Collagen is so successful as a structural protein throughout the animal kingdom, changing little throughout evolution, that it is nearly impossible to identify the animal source of a glue by looking at its amino acid composition alone. The exact sequence of amino acids will determine the animal source and this requires sophisticated and expensive sequencing instrumentation. Recent studies indicate that the fatty acid component (2–10 percent) of animal glues may be used to identify the animal source of animal glue.[20] Analysis by gas chromatography–mass spectroscopy (GC-MS) has revealed fairly distinct ratios of palmitate, stearate, and lesser amounts of azelate in glue samples, in sample sizes comparable to those taken from objects.[21]

Casein, another proteinaceous material, is prepared by denaturing milk proteins with acid and precipitating them out of the solution.[22] Casein contains high amounts of glutamate and proline and a fairly even balance of essential amino acids (see table 1, columns 7–8). These two amino acids—glutamate and proline—are usually sufficient indicators of casein provided no contaminants interfere with the analysis. As we will see, identification of casein can be difficult when calcium-containing pigments or ground layer are present, because calcium can lower the recovery of glutamate and aspartate by as much as 75 percent, perhaps by interfering with the amino acid derivatization step.[23] A more specific marker for identifying casein, as hydroxyproline is for collagen, would help confirm the presence of casein in proteinaceous mixtures and in instances when deleterious components such as calcium-containing pigments hamper the identification of casein.

The protein component of egg is distinguished from casein and animal glue by the 2:1 serine-to-glycine ratio and the slightly higher alanine-to-proline ratio.[24] Of the four binders, egg yolk and egg white are the most difficult to differentiate from each other by amino acid analysis (see table 1, columns 3–6). The even distribution of essential amino acids that makes egg the ideal source of dietary protein is found in both the yolk and the white and makes it difficult to distinguish between them. Furthermore, the composition of egg is close to that of other proteins such as blood serum albumin,[25] which was used in blood glues (see table 1, column 9). Here, too, specific markers for egg yolk and egg white would help distinguish between the two and avoid misidentification with other materials of similar amino acid composition.

Many of the techniques tested during this research sought to distinguish between egg yolk and egg white in one single analysis. At present, two samples must be taken to characterize egg material: one to identify the proteinaceous material with amino acid analysis, and a second to confirm the presence of fats contained in egg yolk but not in egg white. The fatty acid analysis of egg yolk, usually done on a gas chromatograph (GC), will show a small or no azelate peak and a palmitate-to-stearate ratio of approximately 2.5:1, while analysis of drying oils such as linseed, walnut, and poppy oils will show a large azelate component and different palmitate-to-stearate ratios.[26]

Search for Specific Markers for Binders

The first research goal was to find components that might differentiate between egg yolk and egg white and give further confirmation of the presence of casein. Ideally, these markers would unequivocally characterize one material and be lacking in all others, as is the case of hydroxyproline in collagen. For example, in terms of the organic origins of these materials, to enhance a particular protein's cellular function in vivo, sugar groups may be attached to specific amino acid sites on the protein while other amino acids may undergo phosphorylation.[27] I began by looking for specific amino acids that may have been modified with sugars or phosphorous groups.[28]

Table 1 Amino acid analysis results and published compositions of reference materials

	Skin Glue		Egg Yolk		Egg White		Casein		Bovine Serum Albumin	Dammar	Sample of Egg Tempera Paint with Attached Gesso Ground		Glue-Egg Yolk Mixed Medium	
Column Number	1	2	3	4	5	6	7	8	9	10	11	12	13	14
AMINO ACID														
Aspartate	5	5	12	9	13	9	8	5	10	10	5	1	9	11
Glutamate	7	7	11	13	12	14	19	20	15	12	8	3	12	12
Hydroxyproline	10	11?	*	*	*	*	*	*	*	*	10	6	4	3
Serine	4	4	11	9	9	7	7	5	5	12	4	7	9	10
Glycine	32	34	6	3	6	4	4	2	3	23	33	19	17	13
Histidine	1	1	1	3	2	2	2	3	3	1	1	1	2	2
Arginine	5	5	5	7	4	6	3	3	4	4	5	6	5	5
Threonine	2	2	6	6	5	4	5	3	6	3	2	4	4	5
Alanine	11	11	8	5	9	6	4	3	8	5	11	11	8	9
Proline	12	11?	5	4	4	4	12	10	5	2	12	11	7	7
Tyrosine	0	0	3	4	3	4	4	5	3	4	0	3	2	2
Valine	2	2	7	6	7	7	7	6	7	4	2	6	4	6
Methionine	1	1	2	3	4	4	2	8	1	3	1	2	1	1
Isoleucine	1	1	6	6	5	6	5	5	3	4	1	5	3	4
Leucine	2	2	9	9	8	8	8	8	11	8	3	8	6	7
Phenylalanine	2	1	4	5	5	6	4	5	5	2	2	3	2	3
Lysine	2	3	5	7	6	6	6	8	11	3	3	4	4	3

KEY AND NOTES

* Below detectable level or less than 0.5 percent

AMINO ACID COMPOSITION OF REFERENCE STANDARDS

Results are given as mole percent and are derived from both NGA in-house analyses and from published compositions, when available. The most commonly used proteinaceous artists' materials are animal glue (including gelatin, hide, and bone glues), egg yolk, egg white (or glair), and casein. Note the 10 percent hydroxyproline and the 1:8 serine-to-glycine ratio in animal glue, the 20 percent glutamate and 12 percent proline in casein, and the rather balanced compositions of egg yolk and egg white. The published composition of serum albumin has been included because blood glue was often used as a water-resistant glue. The trace amino acid component from dammar resin has been included because it frequently appears in samples; a published composition is not available (the proteinaceous component constitutes approximately 1 percent of the resin). These samples were not analyzed for tryptophan and cysteine

COLUMN 1: In-house analysis of collagen from Sigma, data file 1030A32A (see fig. 4c–d)
COLUMN 2: Composition of rat skin collagen from Herbert A. Sorber, *Handbook of Biochemistry: Selected Data for Molecular Biology*, 2d ed. (Cleveland, 1973), C-282 (percent hydroxyproline not specified)

COLUMN 3: Egg yolk aged two years by natural aging, data file 1030D35A (see fig. 4a–b)
COLUMN 4: Egg yolk composition from *Composition of Foods, Dairy and Egg Products*, U.S. Department of Agriculture, Agricultural Handbook 8-1 (Washington, 1976), item 01–033

COLUMN 5: Egg white aged two years by natural aging, data file 1030C34A (see fig. 2c)
COLUMN 6: Egg white composition from *Composition of Foods, Dairy and Egg Products*, item 01–124

COLUMN 7: Casein from Sigma, data file 1030B33A
COLUMN 8: Composition of parmesan cheese, *Composition of Foods, Dairy and Egg Products*, item 01–125

COLUMN 9: Sequence composition of bovine serum albumin derived from the National Biomedical Research Foundation, Georgetown University Medical Center, Washington, entry ABBOS

COLUMN 10: Dammar resin with 2 percent Tinuvin 292 aged 1,400 hours in fadeometer, supplied by E. René de la Rie, data file 7_09A56A (see fig. 6a–b)

COLUMNS 11–12: Supernatant and precipitate fractions, respectively, from sample of an egg tempera test panel, containing egg yolk paint and gesso ground, with glue from gesso appearing primarily in water-soluble supernatant (fig. 5a [column 11]) and a majority of egg yolk paint appearing in precipitate (fig. 5b [column 12])

COLUMNS 13–14: Supernatant and precipitate fractions, respectively, of hide glue-egg yolk mixed medium, artificially aged 75 hours in a weatherometer, with both compositions reflecting the presence of hide glue and egg yolk (fig. 5c [column 13]–d [column 14])

In fact, egg white and casein do contain a large number of amino sugars and phosphoamino acids. It was hoped that the higher amount of modified amino acids might distinguish egg white from egg yolk, as egg white contains 54 percent ovalbumin, a phosphoglycoprotein,[29] although egg yolk also has a phosphoprotein and glycoprotein component.

Another approach was to look for amino acids such as tryptophan,[30] usually destroyed during the acid hydrolysis, by using an alternate hydrolysis method.[31] The sulfur-containing amino acids, cysteine and methionine, are also highly susceptible to degradation during hydrolysis.[32] Using a performic acid pretreatment, these two amino acids were converted to more stable forms—cysteic acid and methionine sulfone.[33]

Techniques that focused on improved recovery of tryptophan, cysteine, methio-

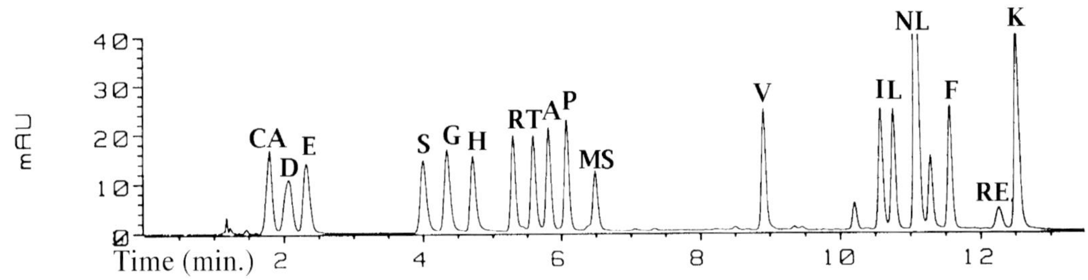

fig. 2a

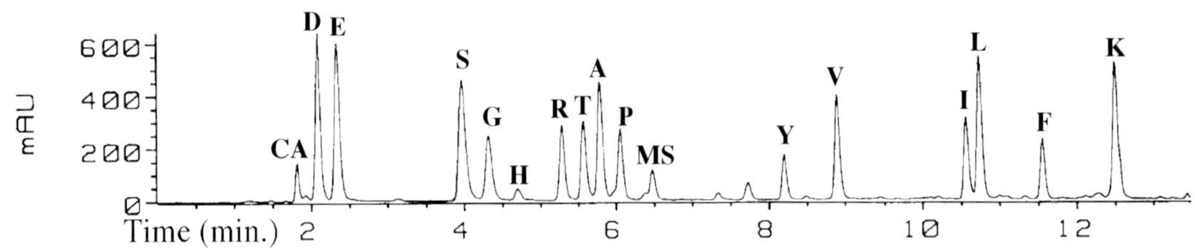

fig. 2b

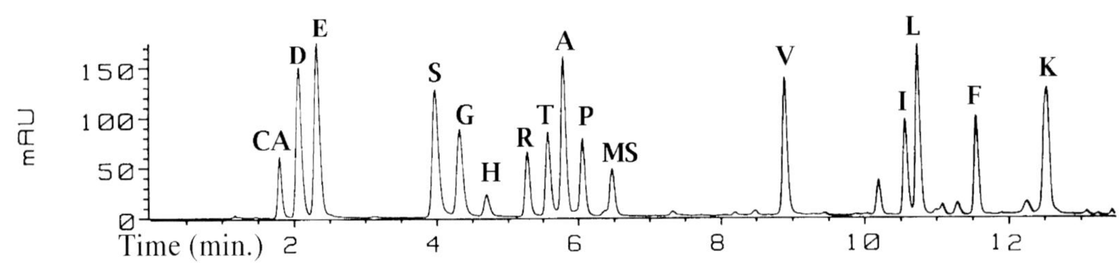

fig. 2c

2. Chromatograms showing the effect of performic acid treatment on (a) amino acid reference standard, (b) egg yolk, and (c) egg white. This treatment is undertaken for an accurate estimate of the amino acids methionine and cysteine, which are converted to methionine sulfone and cysteic acid, respectively. An unwanted side effect is the destruction of tyrosine. However, in tests with reference materials, tyrosine was destroyed in egg white but not in egg yolk, perhaps protected by the fat moiety in the yolk. The protective effect, which would leave tyrosine as a distinguishing marker between egg yolk and egg white, did not seem to hold up when pigments were present in egg yolk samples (for comparison, see fig. 4a–b)

nine, the amino sugars glucosamine and galactosamine, and phospho-amino acids such as phosphoserine were tested on samples of casein, egg yolk, and egg white.[34] See figure 1b for a chromatogram of a reference standard containing the modified amino acids.

The results from the tryptophan, amino sugars, and phosphoserine tests were discouraging because of the low yields of these amino acids from the materials of interest. Associated factors such as the presence of carbohydrates in the samples and the age of the materials may have impeded an adequate recovery of components such as the amino sugars. Monitoring samples for the above-mentioned specific amino acids was not considered worthwhile for routine analyses of artists' materials.

One technique showed promise due to unexpected results. Treating samples with performic acid before hydrolysis produced better recovery in cysteine and methionine but also showed that the levels of these amino acids were fairly similar in both egg yolk and white. However, the amino acid tyrosine, which is usually destroyed during performic acid pretreatment, appeared in the egg yolk sample but not in the egg white (fig. 2a–c). Could this treatment be used to distinguish between egg yolk and egg white? Unfortunately this effect, in which tyrosine was perhaps protected by the fatty components in the egg yolk, did not seem to hold up in samples where pigments were present. The performic acid treatment was continued for approximately six months with actual samples to test its usefulness. As it involved an additional step, seemed to disturb the chromatography by producing unknown peaks, and did not provide any clear advantage in the results, it was subsequently eliminated as a routine technique.

Interference from Pigments

Of concern was the effect that associated materials, especially mineral pigments, might have on the reproducibility of the analytical results.[35] Early analytical results showed interference from copper pigments.[36] A test in which copper and calcium-containing pigments were added separately to reference standards confirmed that copper lowered the yield of all amino acids, possibly during the hydrolysis step, while calcium lowered the recovery of aspartic and glutamic acids (fig. 3a–c). Therefore calcium could interfere with the proper identification of casein, and copper present in samples may reduce the amount of amino acids to background levels, making the results difficult to interpret. To obtain accurate results, it would be necessary to isolate in some way the proteinaceous material from the pigments before undertaking their analysis.

One-Hour Extraction Method

To remove the pigments as effectively and as simply as possible from the proteinaceous material, a short one-hour (hereafter 1h) water-extraction method was developed. Samples from an egg tempera panel with a gesso ground, prepared within the laboratory, were used for the extraction tests.[37] Thirty microliters of commercially bottled HPLC-grade water were added to the sample in the hydrolysis tube. The sample was thoroughly mixed and allowed to stand at room temperature for one hour. To ensure that the larger particles were broken up, the sample was sonicated in an ultrasonic water bath for fifteen minutes before being placed in a small centrifuge at maximum speed for another fifteen minutes. After centrifugation, the soluble proteinaceous material was removed by taking up twenty-five microliters of the supernatant with a syringe. This supernatant was placed in a second hydrolysis tube, evaporated to dryness, and analyzed at the same time as the insoluble material remaining in the precipitate.

Several advantages were immediately apparent from this technique (fig. 4a–d). Because the pigments are removed from the supernatant, a cleaner, more precise analysis of the materials can be achieved. More important, the water-soluble proteinaceous components separated out, leaving the insoluble proteinaceous material behind in the precipitate. This separation could be used to sort out the composition of proteinaceous mixtures encountered with most samples. The supernatant phase is always distinguished from the precipitate by the internal standard because norleucine is water soluble and is easily transferred with the supernatant during the water extraction.[38]

The 1h extraction method proved to be a very simple and useful procedure. A careful comparison between the compositions of

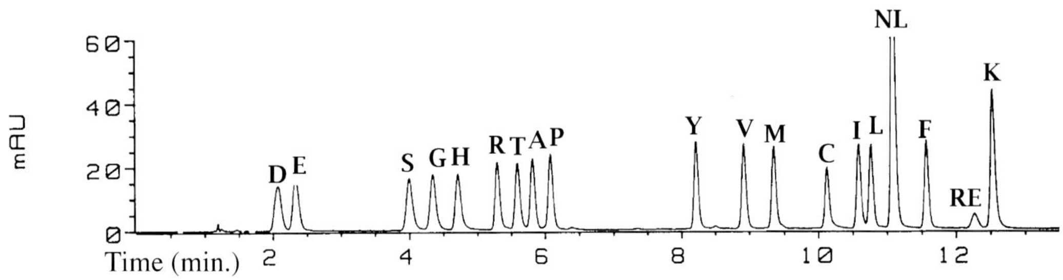

fig. 3a

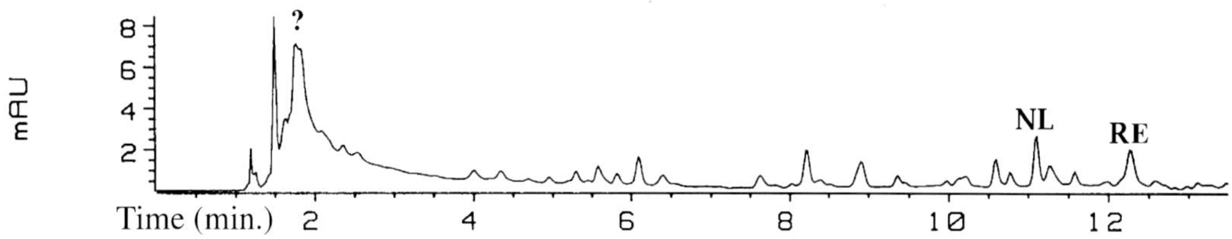

fig. 3b

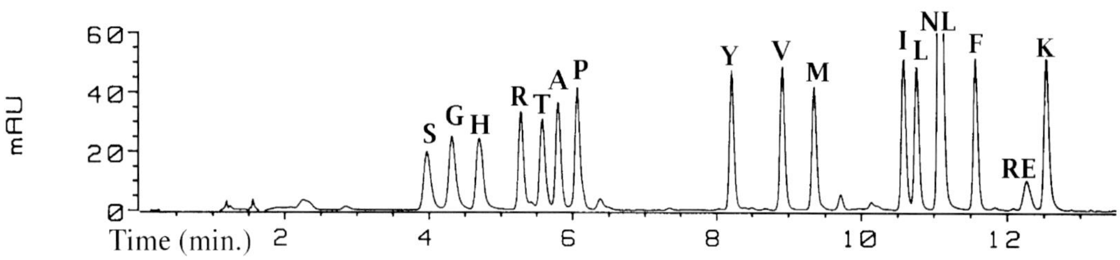

fig. 3c

3. Chromatograms showing effect of (b) copper and (c) calcium on the yield of amino acids. Compare with amounts in reference standard (a) containing 200 picomoles, and note absorbance scale on *y*-axis. The pigmented samples (b and c) have twice the concentration, or 400 picomoles of each amino acid, yet the copper has reduced the yield to trace levels (b). Calcium affects primarily aspartate and glutamate, which elute at approximately 2 minutes on the chromatogram (c)

OPPOSITE PAGE

4. Chromatograms showing 1h extraction of a sample of egg tempera paint and a sample of gesso taken from a panel painting prepared within the laboratory. (a) Supernatant and (b) precipitate of the egg yolk paint layer (see table 1, columns 3–4, for percent composition of egg yolk); (c) supernatant and (d) precipitate of gesso ground layer prepared with rabbit skin glue (see table 1, columns 1–2 for percent composition of collagen). The majority (90 percent by peak area) of the egg yolk paint sample remained water-insoluble in the precipitate (b), while the rabbit skin glue in the gesso sample easily dissolves into the supernatant (c) (40 percent by area). The insoluble pigments remain in the precipitates, allowing a clean analysis of the glue dissolved in the supernatant (c) without deleterious effects of calcium pigment (compare fig. 3c). The egg yolk samples (a) and (b) were treated with performic acid, which affects the tyrosine peak eluting at approximately 8 minutes, converts cysteine to cysteic acid (peak eluting at 1.7 minutes), and converts methionine to methionine-sulfone (peak at 6.5 minutes, coeluting with phenylthiourea peak)

fig. 4a

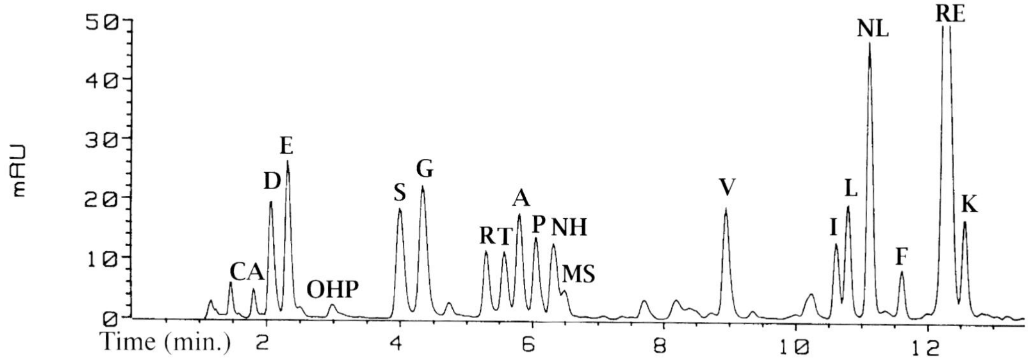

fig. 4b

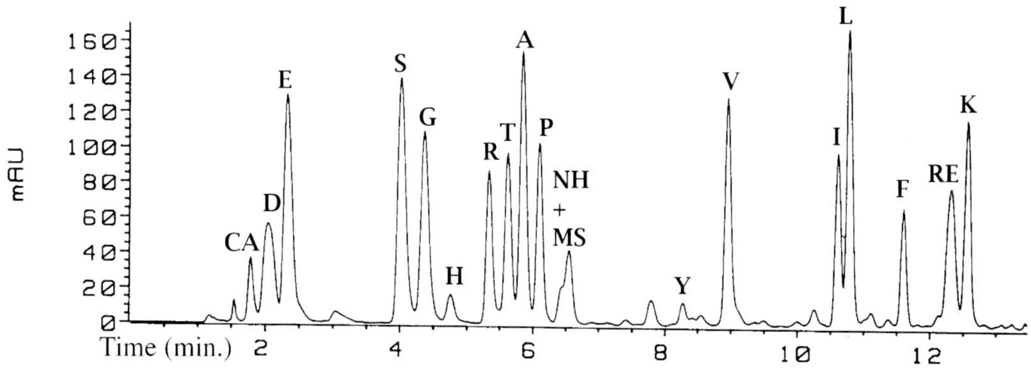

fig. 4c

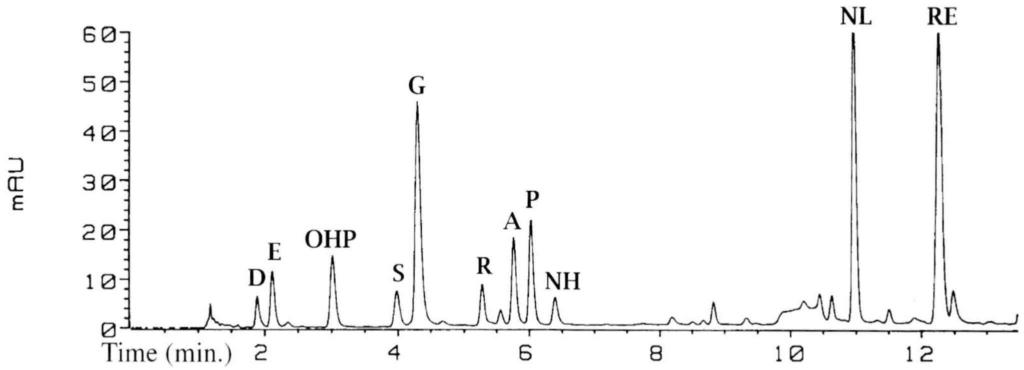

fig. 4d

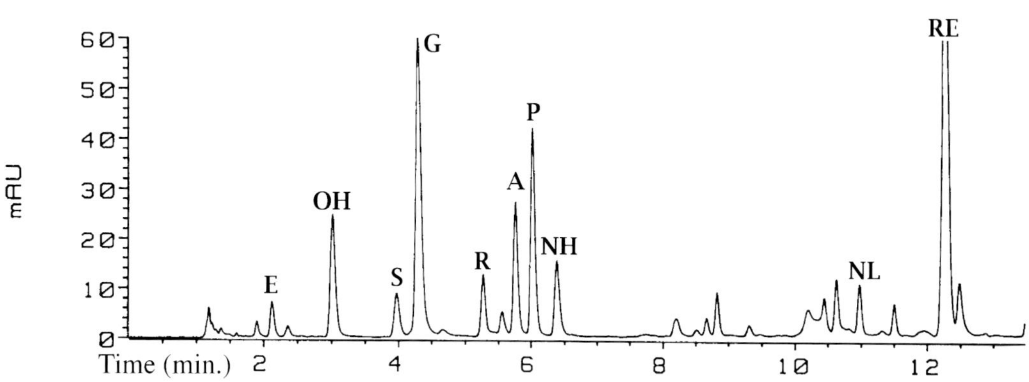

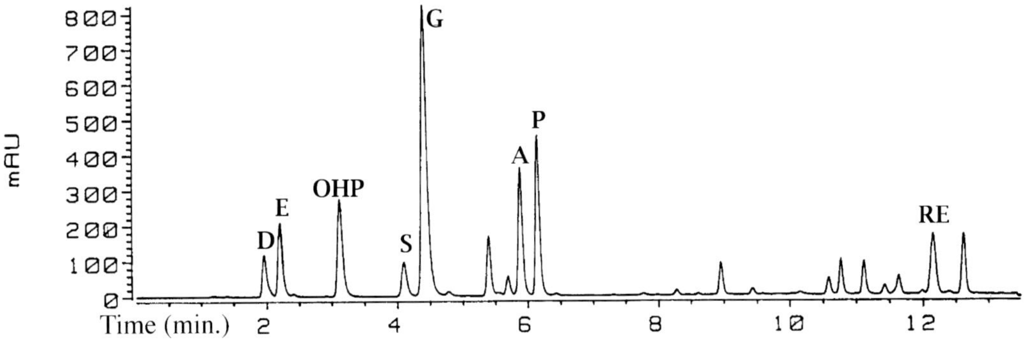

fig. 5a

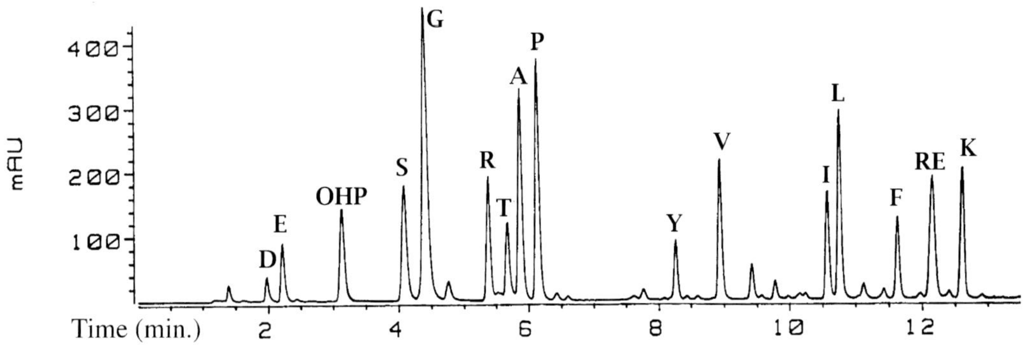

fig. 5b

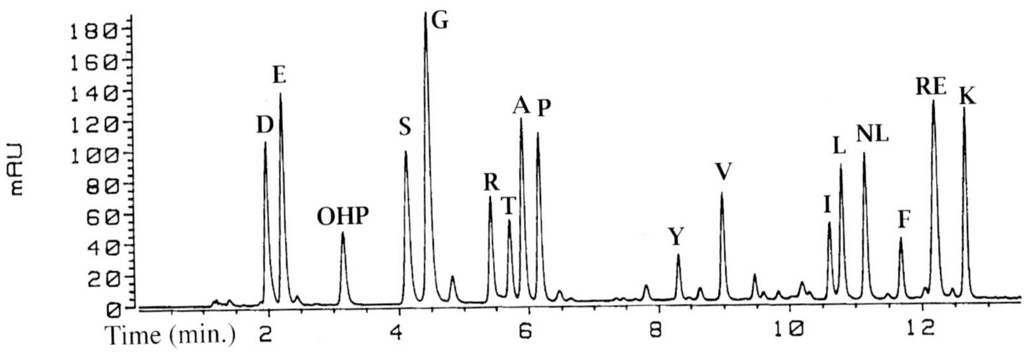

fig. 5c

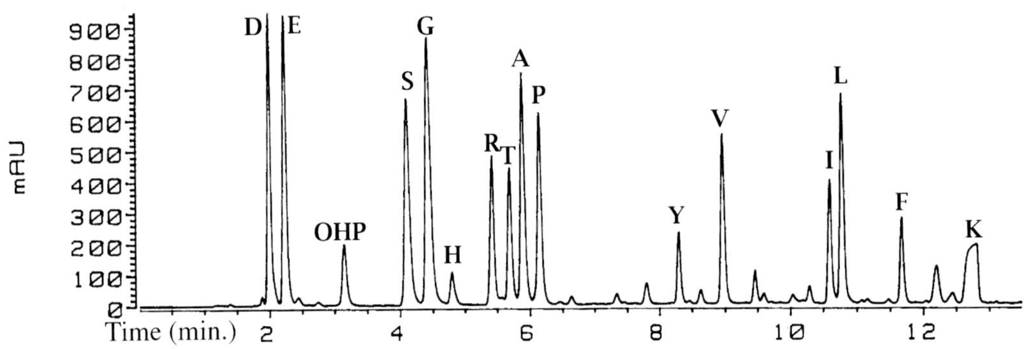

fig. 5d

OPPOSITE PAGE
5. Chromatograms showing solubility differences after the 1h extraction between two egg yolk and animal glue preparations. An egg tempera preparation, that is, egg yolk prepared in a paint layer over a rabbit skin glue gesso ground, yields only dissolved glue from the gesso in the (a) supernatant fraction, while egg yolk and glue appear in the (b) precipitate fraction (see table 1, columns 11–12 for percent composition). A mixed medium of egg yolk and animal glue mixed 1:1 and artificially aged as a film 75 hours in a weatherometer yields approximately the same ratio of egg yolk and glue in both the (c) supernatant and the (d) precipitate fractions (see table 1, columns 13–14 for percent composition)

the water-soluble components in the supernatant and the insoluble components in the precipitate of a given sample could indicate structural information within the sample. For example, if animal glue is present in a layer adjacent to an egg yolk paint surface as a ground layer, as a distinct paint layer, or as a consolidant, a majority of the glue component will remain soluble and appear in the supernatant, while a majority of the egg yolk component will remain insoluble and appear in the precipitate (fig. 5a–b, and table 1, columns 11–12). However, if the glue is intermixed with egg yolk before application, approximately the same ratio of glue-to-egg yolk will appear in both the supernatant and the precipitate (fig. 5c–d, and table 1, columns 13–14). The egg yolk seems to bind with a larger portion of glue when mixed before drying, leaving the glue less water soluble. A glue relining of an egg tempera painting will also bind the glue to the egg layer to some extent, since during relining the glue is forced, using heat, through the paint surface. Of course extraneous factors, such as the amount of surface area of the sample, may cause a degree of variation in the solubility of the materials.

With the 1h extraction method we begin to gain insight into the structure and treatment of the painting, which can then be clarified and confirmed by cross-sectional analysis.

Nonproteinaceous Materials

To avoid misinterpretations, a number of nonproteinaceous materials were analyzed. These materials included linseed, poppy, and walnut oils; dammar resin; mastic; rosin; wheat paste; gum arabic; and gum tragacanth.[39]

The samples were analyzed with and without the PITC derivatization step to identify un-derivatized non-amino acid components that absorb at 254 nanometers and that would mask PTC–amino acid peaks. Any peaks appearing in the un-derivatized analysis were assumed not to be amino acids and subtracted from the derivatized results. For example, an unidentified non-amino acid peak, which coelutes with the derivatized amino acid valine (at approximately 9 minutes) appears in samples of oils, resins, and gums. The peak can be distinguished from the amino acid by comparing the spectrum to that of PTC-valine. Furthermore, the peak displays a different spectrum depending on whether it is found in an oil, a resin, or a gum.[40] This peak should be noted, both to correct the percent composition of valine and to indicate the presence of a nonproteinaceous material in the sample.

The analytical results were surprising as they revealed trace amino acids in almost all the nonproteinaceous materials. Wheat starch, dammar resin, gum arabic, and gum tragacanth contain trace amounts at levels and patterns possibly unique enough to characterize samples. In dammar, the proteinaceous component represents approximately 1 percent by weight of fresh resin (table 1, column 10, and fig. 6a–b), and the pattern has been discerned in a number of actual samples containing triterpenoids, examples of which will be discussed later. Actual paint samples containing gums and starches have yet to be analyzed; therefore, it is not known if the sample size normally taken from paintings will be sufficient to indicate these materials. Trace amino acids should be used only as indicators, however, and other analytical methods such as carbohydrate and oil-resin analysis should be used to identify these materials.

The cold-pressed linseed and walnut oils showed a higher amino acid content than did the heat-extracted or boiled oils, but it is not known if these amounts are sufficient for identification in actual samples. The nine-minute peak also showed a spectrum for the cold-extracted oils different from that for the boiled and heat-extracted oils. Fresh mastic resin seemed to contain trace amino acids as well; the pattern was different from that of dammar, but the amount recovered was lower,

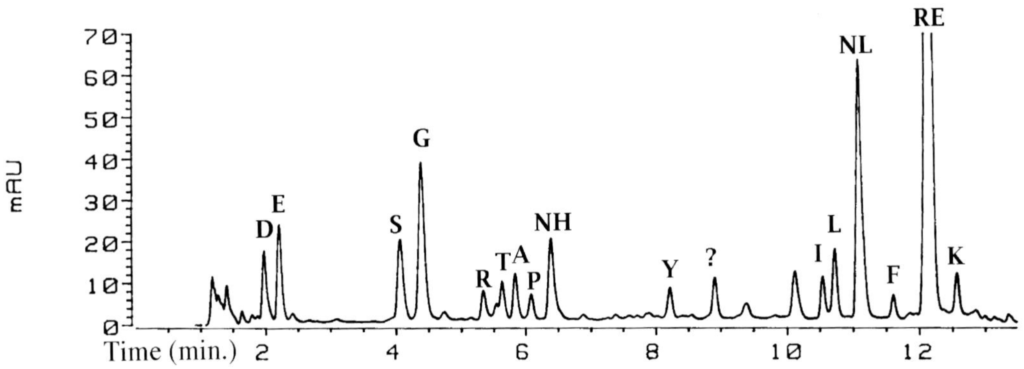

fig. 6a

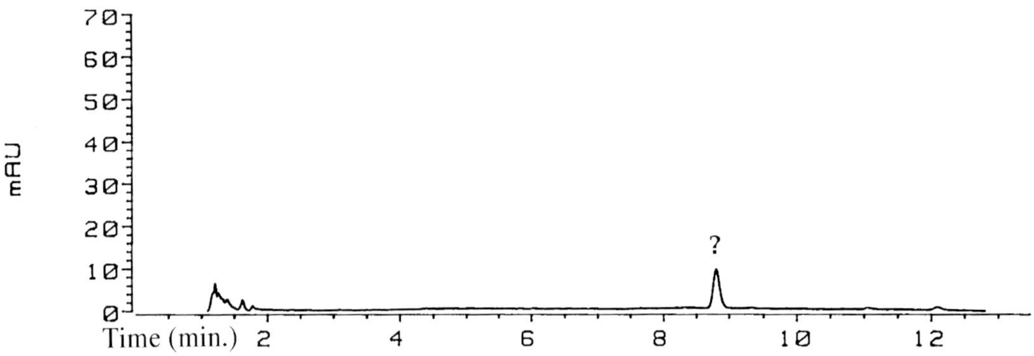

fig. 6b

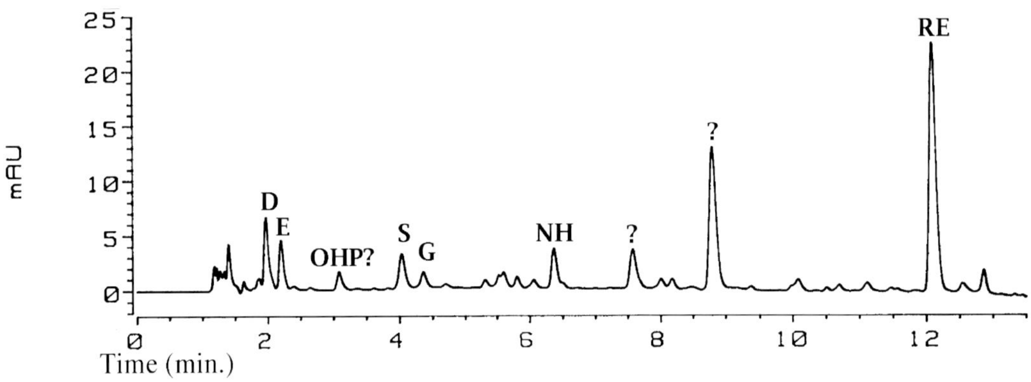

fig. 6c

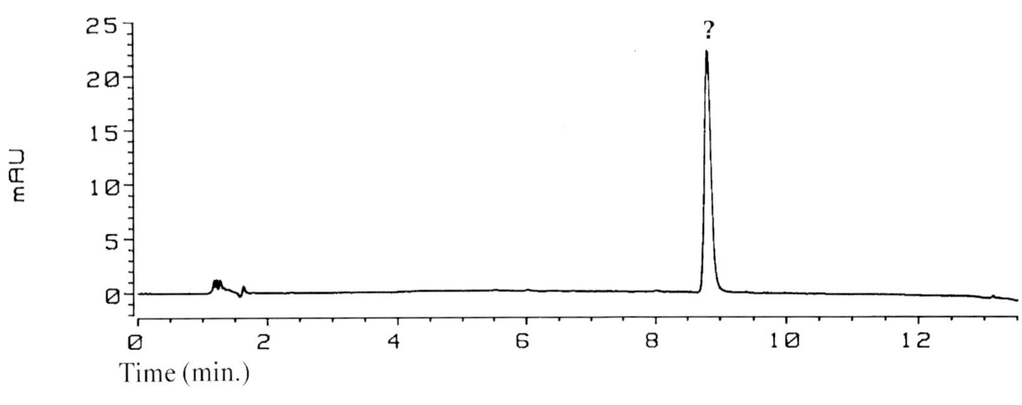

fig. 6d

OPPOSITE PAGE
6. Chromatograms showing trace amino acids found in dammar and mastic resins. (a) Derivatized and (b) underivatized samples of dammar, showing that the derivatized peaks in (a) are consistent with derivatized amino acids (table 1, column 10), and are not nonproteinaceous components absorbing at 254 nanometers. The supernatant and precipitate fractions were added together in chromatogram (a). The dammar sample has been artificially aged 1,400 hours in a fadeometer and has 2 percent Tinuvin 292 added as an ultraviolet absorber. The same pattern is seen in fresh dammar samples, but extraction problems and interference from other components in fresh dammar make this aged sample more characteristic of the trace pattern seen in actual samples. Sample was prepared by E. René de la Rie, while at the Metropolitan Museum of Art, New York. (c) Derivatized and (d) underivatized samples of unaged mastic resin pellets from Kremer-Pigmente. Trace amino acids in mastic show a pattern different from those in dammar, especially in the serine-to-glycine ratio. The low amino acid levels are caused either by the insolubility of fresh material, lower absolute amounts, or both. The mastic sample was not extracted into two fractions

perhaps because the insolubility of the fresh material in the water-based sample preparation did not allow proper derivatization of the proteinaceous component. Rosin did not show any amino acid content at all, possibly for the same reason given for mastic. Miscellaneous substances such as human fingerprints, papyrus, and garlic and onion juice also showed a significant amino acid content, although the amounts of these substances in samples from actual objects may be too small for identification.

Results of Amino Acid Analysis Using the 1h Extraction Procedure

The outcome from the above tests of reference materials was indispensable for the interpretation of results from actual samples and provides insight into the variations that can be expected in the results.

What follows are brief accounts of technical inquiries into specific paintings with discussions on the analytical results. To answer questions from conservators regarding objects undergoing treatment, small samples from the object were submitted for amino acid analysis when proteinaceous materials were thought to be present. The 1h extraction procedure was used on all samples. Some samples were also pretreated with performic acid to ascertain whether the tyrosine peak would mark the presence of egg yolk. After adding the internal standard to monitor amounts, the samples were extracted using the 1h extraction method, hydrolyzed, derivatized, and analyzed according to the procedure described above.

The results given below were selected to demonstrate a variety of materials and techniques analyzed from paintings in the National Gallery of Art. The paintings ranged from Italian tempera panels and a Renaissance canvas painting to a pre-Raphaelite painting on paper. Included are the analyses of paint binders and a ground layer, as well as various supports. The results are given chronologically with works by Gaddi of the early Florentine Renaissance, Mantegna of fifteenth-century Ferrara, the pre-Raphaelite Burne-Jones of nineteenth-century England, and Gauguin and Toulouse-Lautrec of late-nineteenth-century France.

Agnolo Gaddi, *Madonna Enthroned with Saints and Angels*

The materials used in Gaddi's fourteenth-century Florentine triptych, *Madonna Enthroned with Saints and Angels* (fig. 7), were investigated as part of a comprehensive study of the artist's methods and materials. This panel painting is important not only because of its technical mastery and well-preserved condition but also because Agnolo was the son of Taddeo Gaddi, who was a pupil of Giotto. Agnolo was also the teacher of Cennino Cennini.[41] Careful analysis of Gaddi's work could help place the methods described in Cennini's classic artist manual, *Il Libro dell'Arte*, in a better-informed context.

Media analyses indicated that Gaddi may have used different binders on the same panel depending on the color he was applying.[42] For example, egg tempera seemed to have been used for some colors, and glue distemper for others. Layers of both and possibly a mixture of the two were used for yet other colors (table 2, columns 1–4, and fig. 8a–h). The analyses were often prompted by the conservator's noticing unusual solubility associated with certain colored areas, which may indicate dissimilar materials.[43]

Two brocade areas painted over gold leaf were analyzed. The conservator described the white brocade of the madonna's tunic as extremely water insoluble and the vermilion paint in the red floor tapestry brocade below

Table 2 Amino acid analysis results from paintings by Gaddi (columns 1–4), Gauguin (column 5), Burne-Jones (columns 6–8), and Toulouse-Lautrec (column 9)

	Agnolo Gaddi								Paul Gauguin		Edward Coley Burne-Jones						Henri Toulouse-Lautrec	
	Red Brocade		White Brocade		Light Green		Yellow		Gesso Ground		Paper Sizing+		Blue Sky+		White Sky+		Cardboard	
Column Number	1		2		3		4		5		6		7		8		9	
AMINO ACID	S	P	S	P	S	P	S	P	S	P	S	P	S	P	S	P	S	P
Aspartate	4	11	16	17	9	6	8	8	6	2	8	7	7	6	10	9	13	8
Glutamate	12	13	12	13	9	7	11	13	8	3	8	9	7	8	9	10	10	12
Hydroxyproline	9	5	*	*	6	2	4	4	9	9	4	8	*	4	*	1	3	2
Serine	5	9	8	10	6	10	10	9	4	5	6	4	7	7	8	10	7	9
Glycine	31	17	15	12	30	14	26	22	36	33	19	31	8	16	14	17	24	18
Histidine	*	*	2	1	1	1	1	1	1	1	3	1	4	3	2	2	3	1
Arginine	3	3	3	4	3	3	4	4	4	5	5	2	6	4	6	6	1	1
Threonine	2	4	5	5	2	8	4	4	2	3	4	2	6	4	5	5	3	5
Alanine	11	9	10	8	12	12	8	8	12	13	7	10	6	7	7	7	12	14
Proline	9	8	4	4	8	10	7	8	10	13	8	10	6	10	6	6	5	8
Tyrosine	1++	1++	1	1	1	3	2	2	*	1	1++	1++	1++	3++	*	*	1	1
Valine	3	5	7	6	3	9	4	4	3	4	4	4	10 ‡	4	‡	‡	5	9
Methionine	2	2	2	1	*	1	1	1	*	1	3	1+	3	2+	3	3+	1	2
Isoleucine	1	2	5	5	3	5	3	3	1	2	6	2	6	5	8	4	3	3
Leucine	2	4	7	7	4	9	5	6	2	4	4	3	6	5	8	6	5	7
Phenylalanine	1	2	2	3	2	3	2	3	1	2	4	2	5	3	5	4	2	2
Lysine	3	*	2	3	2	*	2	2	1	1	3	2	5	4	5	5	2	1
⊗ Percent of fraction	50	50	30	70	30	70	25	75	75	25	20	80	60	40	45	55	20	80
⊙ Media sample, in micrograms	2.9		0.6		1.6		0.7		4.7		1.6		0.7		0.2		2.9	

KEY AND NOTES

S=Supernatant
P=Precipitate
* Below detectable level or less than 0.5 percent
+ Sample was treated with performic acid, after which methionine is converted to methionine sulfone, cysteine is converted to cysteic acid (not shown), and tyrosine is destroyed
++ Possibly tyrosine preserved in egg yolk sample, or possibly unidentified material with same elution (see fig. 2)
‡ Valine peak was overwhelmed by nonproteinaceous material eluting at approximately 9 minutes
⊗ Percent of fraction is based on total peak areas for each chromatogram
⊙ Total media sample size (approximate proteinaceous yield) based on peak areas compared with a reference standard

AMINO ACID COMPOSITION OF SAMPLES DISCUSSED, GIVEN AS MOLE PERCENT

Results from the water-soluble supernatant fraction are followed by the water-insoluble precipitate fraction for each sample

Samples from Agnolo Gaddi, *Madonna Enthroned with Saints and Angels* (see fig. 7)
COLUMN 1: Red paint brocade from tapestry, bottom of center panel (see fig. 8a–b)
COLUMN 2: White paint from brocade from the madonna's tunic, center panel (see fig. 8c–d)
COLUMN 3: Light green paint from Saint Andrew's robe, left panel (see fig. 8e–f)
COLUMN 4: Yellow paint from Saint Catherine's robe, right panel (see fig. 8g–h)

Sample from Paul Gauguin, *Fatata te Miti* (see fig. 15)
COLUMN 5: Gesso ground layer (see fig. 16a–b)

Samples from Edward Coley Burne-Jones, *Saint Barbara* (see fig. 17)
COLUMN 6: Sizing in paper support (see fig. 18a–b)
COLUMN 7: Blue paint from sky (see fig. 18c–d)
COLUMN 8: White paint from sky (see fig. 18e–f)

Sample from Henri de Toulouse-Lautrec, *Maxime Dethomas* (see fig. 19)
COLUMN 9: Sizing in cardboard support (see fig. 21a–b)

7. Agnolo Gaddi, *Madonna Enthroned with Saints and Angels*, c. 1380/1390, panel painting, triptych, tempera and distemper on wooden panel. The figures represented from left to right are Saint Andrew, Saint Benedict, Madonna and Child, Saint Bernard, and Saint Catherine of Alexandria
National Gallery of Art, Washington, Andrew W. Mellon Collection 1937.1.4 a,b,c

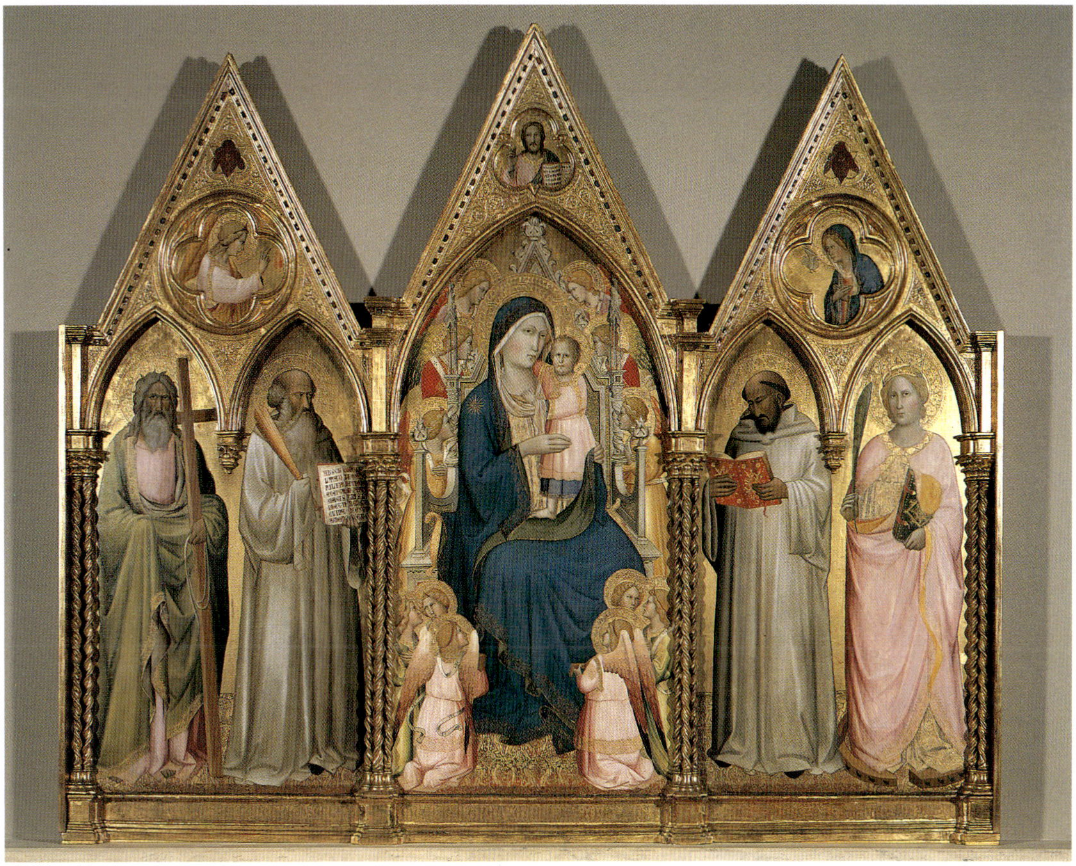

the madonna as very water sensitive yet remarkably well preserved. Amino acid analysis results of the white brocade area showed egg binder (table 2, column 2, and fig. 8c–d). The same analysis of a sample from the red brocade area, however, showed a large amount of animal glue, predominantly in the soluble supernatant, and a small amount of egg material in the precipitate (table 2, column 1, and fig. 8a–b). These solubility characteristics indicated that perhaps the vermilion was applied with glue over gold and coated with a thin coat of egg yolk.[44] The sample was taken from above the gold leaf, so there was no glue interference from the gesso ground or bole layer. Both brocade areas are in very good condition and do not seem to have required a consolidant. The sample was analyzed for gums, and the absence of significant amounts of sugars indicated that gums were not present.[45]

A cross section (fig. 9) of the red brocade area clearly shows a thick orange layer over the gold leaf and a thin, darker red paint layer covering it.[46] This cross section corresponds with the glue-vermilion layer covered by a thin egg–red lake glaze. In other areas the red glaze seems to have been pigmented with red lead (fig. 10), depending on the effect required.

In the section "How to Design Brocades in Various Colors," Cennini recommended, "Then, if you want to make the cloth red, lay in with vermilion over this burnished gold. If you need to put any dark on it, put it on with lac; if you need to put on lights, put them on with red lead; all tempered with yolk of egg."[47] Gaddi seems to have used this technique to tone the receding area of the red brocade tapestry with red lake and the highlights, such as the tops of the steps, with red lead. Although Cennini did not specifically mention the use of glue as a binder for the vermilion, his reference to egg yolk as the binder ("all tempered with yolk of egg"), describing the toning glazes infers a choice of media, if not a definite recommendation of egg yolk. For example, in blue brocades Cennini recommended a lead white layer tempered in egg yolk, covered by ultramarine in a glue-egg mixture.[48] Furthermore, in his section on vermilion, he mentioned, "This color calls for various temperas, according to the situations in which you have to use it."[49] Glue was certainly an option as a binder for vermilion. The application of a lake pigment glaze with egg yolk to modulate an underlying red layer has been noted in the work of the fourteenth-century Florentine painter Lorenzo Monaco.[50] In addition to the glazing effect, the upper coat of egg yolk may also have offered a certain amount of surface protection to the more fragile and hygroscopic glue layer below and have unified the surface texture of the brocade with the surface of the primarily egg tempera painting.

Another reason for a red glaze over vermilion may have been as a "filter" to protect the vermilion layer from turning black. Robert Feller, investigating the darkening of vermilion, found that it darkened less in watercolor than in oil and that a red filter offered the best protection against the light-induced darkening.[51] His research substantiated the claim that red lake glazes would protect vermilion underpainting from darkening. In describing vermilion, Cennini wrote, "But bear in mind that it is not its nature to be exposed to the air . . . because, in the course of time, from exposure to the air, it turns

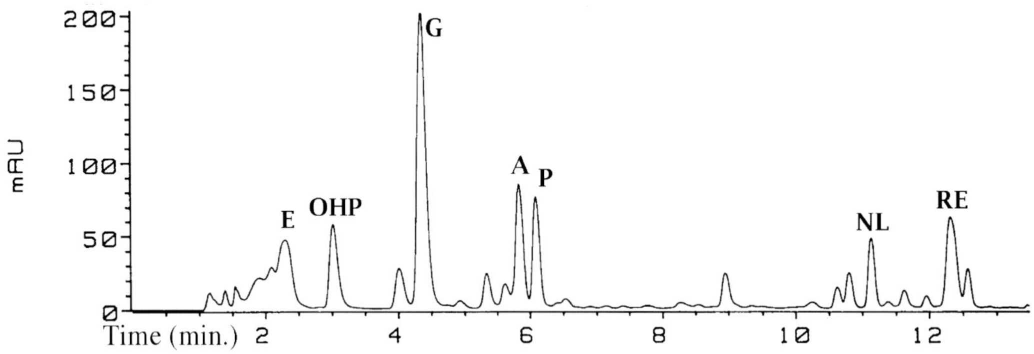

fig. 8a

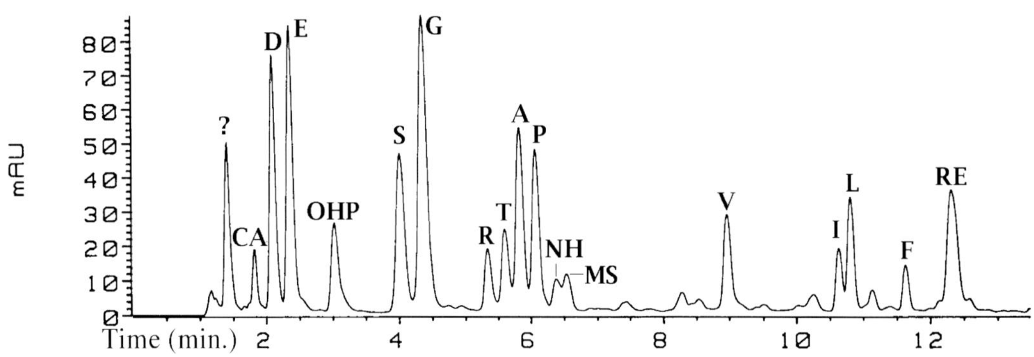

fig. 8b

8. Chromatograms of four samples from Gaddi, *Madonna Enthroned with Saints and Angels* (fig. 7): (a) Red brocade paint, supernatant fraction, containing the majority of the sample material and matching the composition of glue (compare hydroxyproline, serine, and glycine peaks of glue sample in fig. 4c–d) (chromatographic perturbations at 2 minutes do not affect interpretation); (b) red brocade paint, precipitate fraction, with approximately one-fifth of the total sample showing the presence of glue (hydroxyproline is present), but with higher amounts of essential amino acids eluting later in the chromatogram. The composition and insolubility of the precipitate suggest a separate layer of egg yolk adjacent to a large glue-bound layer. Red brocade samples were treated with performic acid

OPPOSITE PAGE
c) White brocade paint, supernatant fraction, showing a low yield and an amino acid pattern consistent with dammar resin (compare to fig. 6a); (d) white brocade paint, precipitate fraction, showing an amino acid composition and water solubility consistent with egg yolk (compare to fig. 4b)

(e) Light green paint, supernatant fraction, with trace amounts of animal glue, possibly from gesso ground below; (f) light green paint, precipitate fraction, showing a composition and solubility consistent with egg yolk

fig. 8c

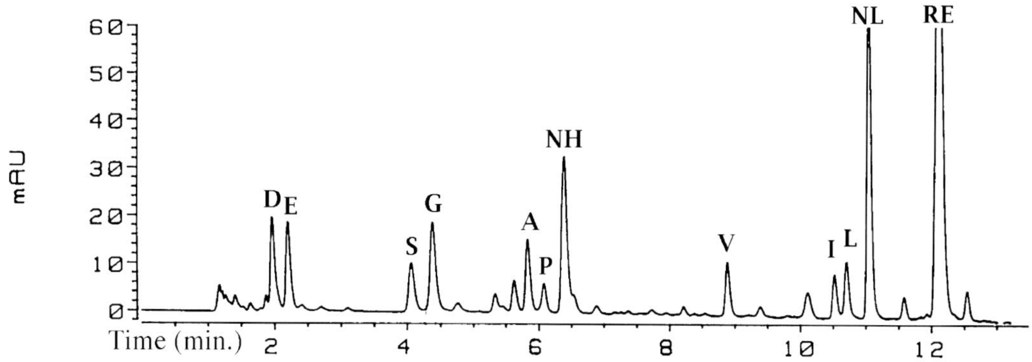

fig. 8d

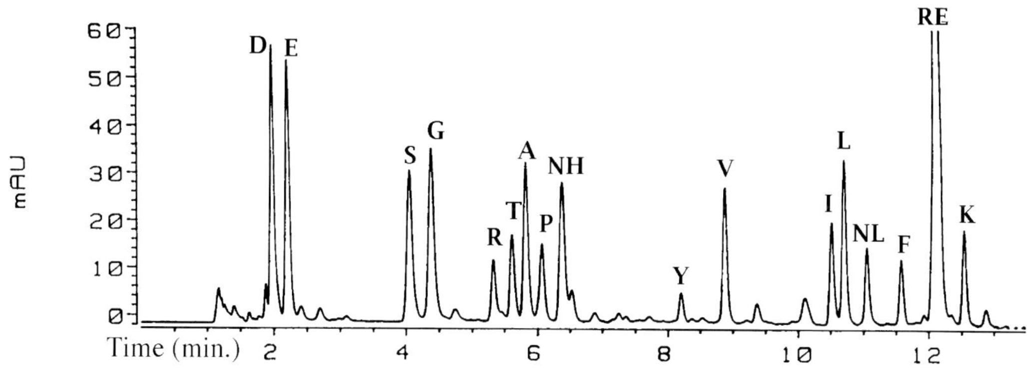

fig. 8e

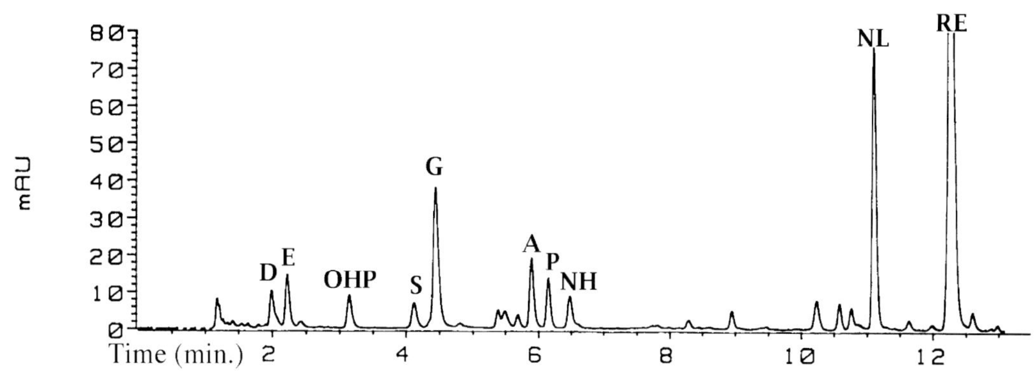

fig. 8f

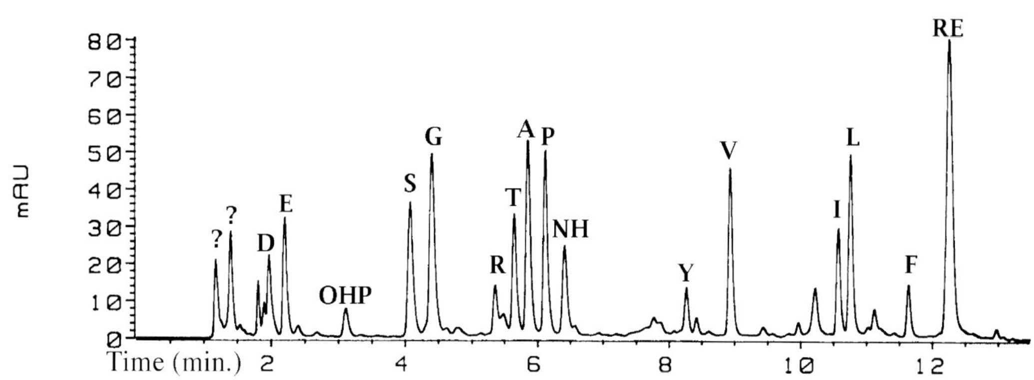

black when it is used."[52] As Cennini's master, Gaddi may have been well aware of the darkening of vermilion over time and used the glazes for a dual purpose of toning and protecting the vermilion.

A test panel with different binder layers was prepared in the scientific research department at the National Gallery of Art. The glue underlayer gave greater control in handling of the sgraffito and produced smoother lines, which may have been another advantage in the gold leaf–animal glue–egg yolk stratification.[53] Further analysis is needed to determine how widespread the use of this technique was in Italian tempera panels.

Analytical results from other painted areas (sampled above the gesso ground layer) also suggested a careful choice of paint binder, depending on the color being used and the nature of the support. The sample of vermilion paint from the angel wings (not illustrated) applied on a gesso ground layer appears to have been painted with egg tempera, whereas in the brocade area, vermilion (table 2, column 1, and fig. 8a–b) was applied with glue. An angel's flesh color (not illustrated), the light green of Saint Andrew's robe (table 2, column 3, and fig. 8e–f), and the light blue of the Christ child's robe (not illustrated) also seem to be painted with egg. There is some small amount of animal glue present in the water-soluble fractions, but the precipitate contains the majority of the sample and matches the composition of egg.

The yellow paint from Saint Catherine's robe was also noted by the conservator as being somewhat water soluble. Amino acid analysis results showed an almost identical composition for the supernatant and the precipitate, with animal glue and egg appearing in similar ratios in both phases (table 2, column 4, and fig. 8g–h). This finding suggests

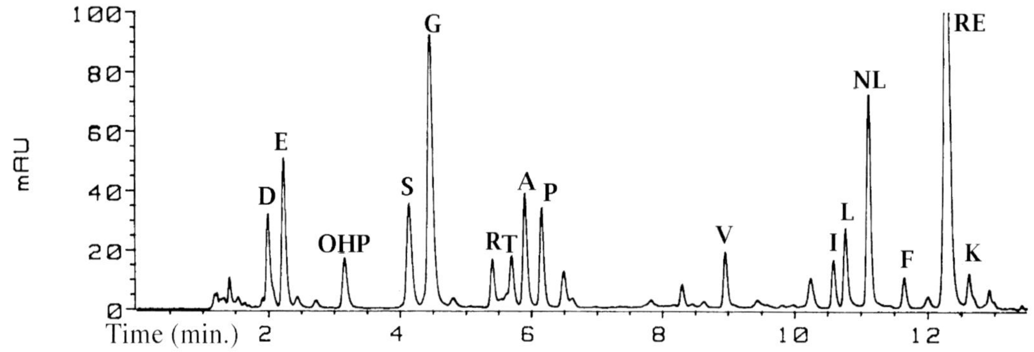

fig. 8g

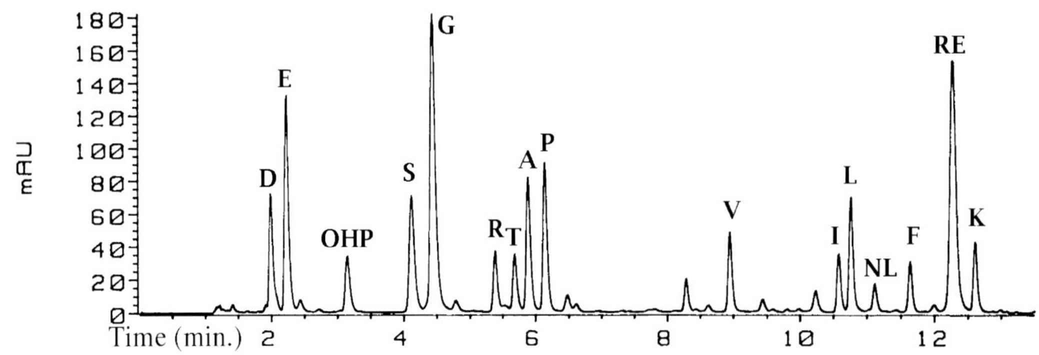

fig. 8h

8. (g) Yellow paint, supernatant fraction, with approximately one-third of the sample material; (h) yellow paint, precipitate fraction. Both the supernatant and the precipitate fraction of yellow paint show a remarkably similar mixture of animal glue and egg (yolk), suggesting that these binders may have been mixed before use (compare to fig. 5c and 5d). The yellow paint was noted by the conservator as being more water soluble than the other painted areas, despite its containing lead white as a pigment. A glue-egg mixture is consistent with the increased solubility of the yellow area

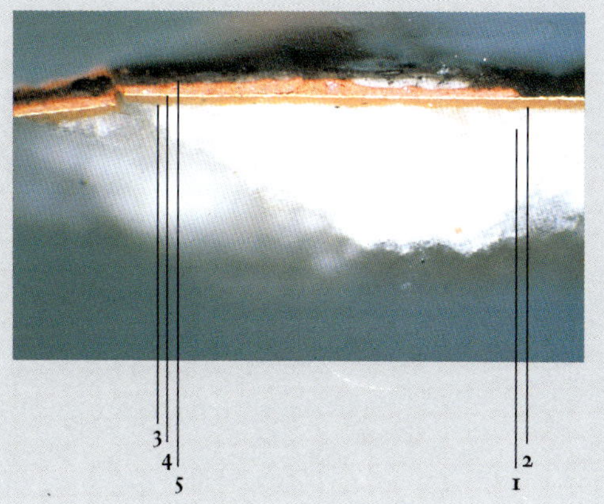

9. Cross section of red brocade paint from Gaddi, *Madonna Enthroned with Saints and Angels* (fig. 7), showing the (5) egg tempera–red lake glaze covering the (4) glue-vermilion paint layer above the (3) gold leaf layer (see fig. 10). Below the gold leaf are the (2) bole and (1) gesso ground layers. The sample was taken from an edge of the brocade, with paint from Saint Bernard's robe (right side panel) brushed over the brocade area and protecting the brocade's top layers during sample preparation

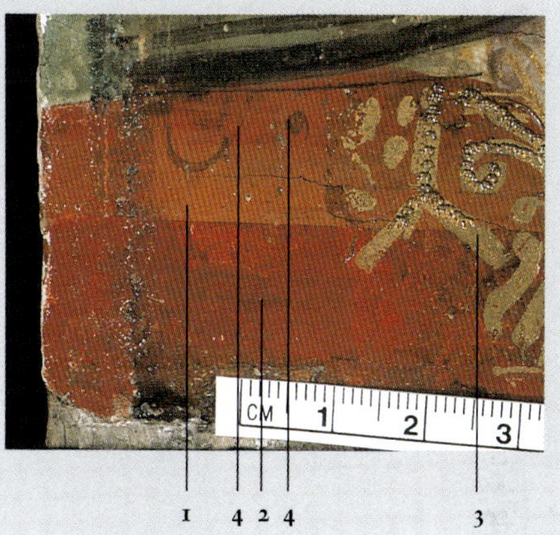

10. Detail from Gaddi, *Madonna Enthroned with Saints and Angels* (fig. 7), center panel, lower left-hand corner, showing area of red brocade protected under frame. Brocade consists of gold leaf covered by overlapping layers of red paint. Note (1) orange-red paint stripe applied over (2) a deep red before (3) sgraffito work. Note also (4) the markings consistent with the extension of the brocade pattern

that the egg was mixed 1:1 with glue for the application of the yellow color. A cross-sectional sample taken from the yellow area indicates a homogeneous layer of paint (not illustrated), rather than several layers with different binders, and this single layer was probably tempered with a mixed medium.

To summarize, medium analysis indicated considerable variation in Gaddi's choice of binder, far more than expected for a tempera panel painting. The red brocade above the gilding appears to have been rendered with animal skin glue and coated with a thin egg layer, possibly to modulate the color and to improve the sgraffito handling, while the white brocade showed egg alone. The vermilion, light blue, light green, and flesh-colored paint appear to be mixed with egg yolk alone. The yellow paint appears to be a 1:1 mixed medium of egg yolk and animal glue. There is some evidence that the madonna's blue robe may have consisted of several paint layers, but damage from a previous restoration made the analytical results unreliable.

Cennini approached pigment binders with great care in his artists' manual and gave specific recommendations for each pigment: "To approach the glory of the profession step by step, let us come to the working up of the colors . . . which one calls for one tempera, which requires another; and just as they differ in their colors, so do they also in the characters of their temperas and their working up."[54] Cennini suggested glue for mixing a number of colors for panel painting. For example, he recommended glue for dark blue pigments, greens, and yellows, while egg yolk was suggested specifically for certain greens and in general for painting on panels. Some colors such as vermilion and azurite could

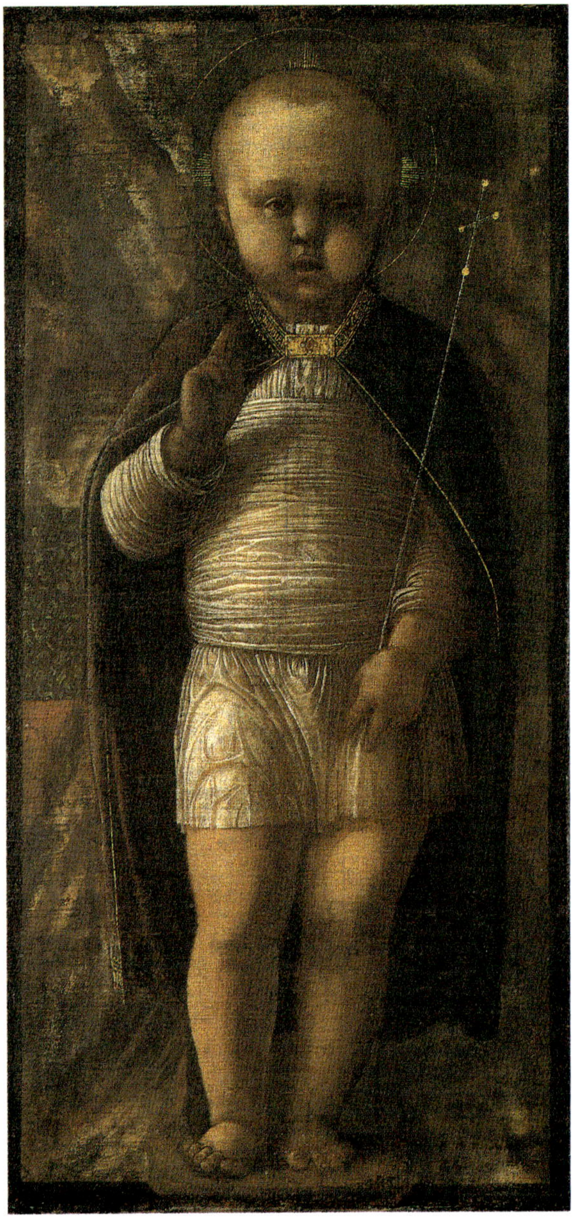

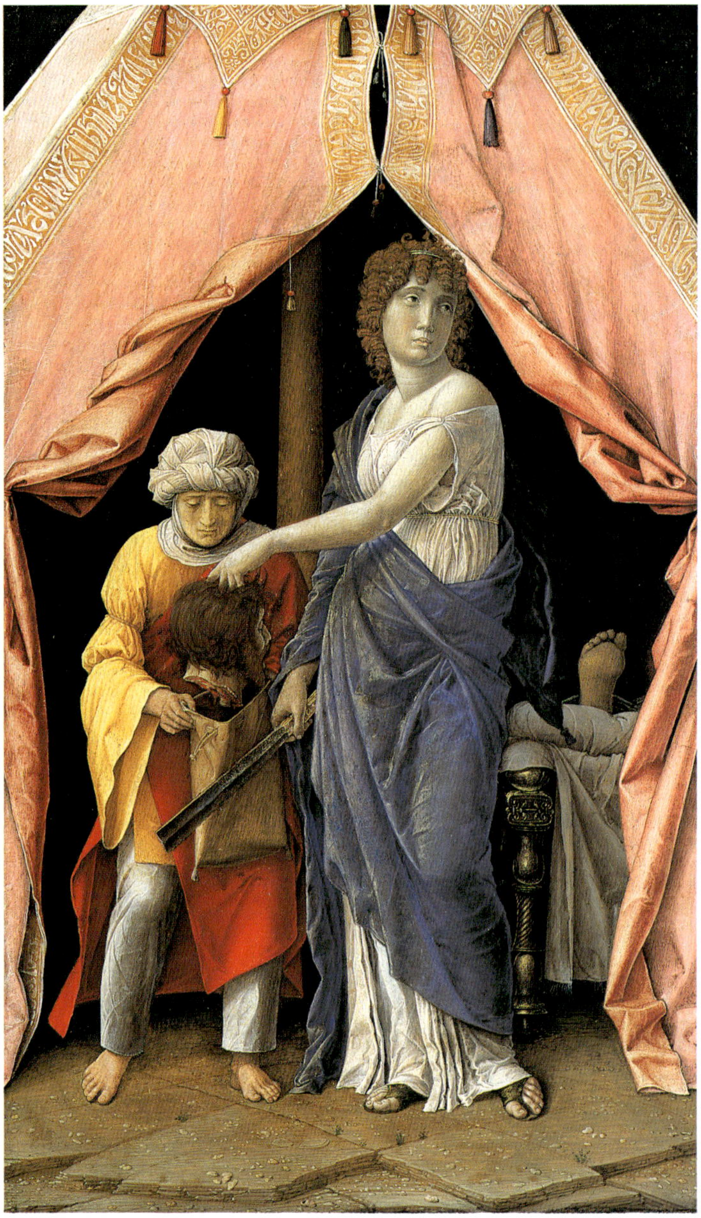

"call for various temperas"; azurite required size if used as a dark blue colorant, but egg yolk when mixed with giallorino (yellow) for a green colorant.[55]

Further analytical work should reveal whether mixed media on panel painting was a common practice in the early Renaissance and whether the selection of glue for mixing certain colors was universally practiced or embraced by only certain artists. The selection of different binders implies that the artists gave far more care to the actual application of the paint than was previously thought.

Andrea Mantegna, *Christ Child Blessing* and *Judith and Holofernes*

Results of media analyses from panel paintings by Gaddi, described above, and Cosimo Tura[56] indicate that different paint binders were applied for specific colors on the same panel. Paint from Mantegna's canvas painting, *Christ Child Blessing* (fig. 11), and panel painting, *Judith and Holofernes* (fig. 12),[57] were analyzed in order to see if this artist also selected different binders for use with specific colors.[58] The canvas painting was thought to have been rendered with distemper, and the panel with egg tempera.

11. Andrea Mantegna, *Christ Child Blessing*, c. 1480/1490, egg tempera on canvas. Note that the black edge along the perimeter of the painting, found on many fabric paintings, has been broken at the bottom by the platform supporting the Christ Child and extending toward the viewer
National Gallery of Art, Washington, Samuel H. Kress Collection

12. Andrea Mantegna, *Judith and Holofernes*, c. 1495, egg tempera on wooden panel
National Gallery of Art, Washington, Widener Collection

Table 3 Amino acid analysis results of samples taken from Mantegna, *Christ Child Blessing* and *Judith and Holofernes*

	Christ Child Blessing			*Judith and Holofernes*		
	Off-white Paint+		Precipitate Minus Normalized Animal Glue Composition	Off-white Paint+		Precipitate Minus Normalized Dammar Composition
Column Number	1		2	3		4
AMINO ACID	S	P	P -N=	S	P	P -N=
Aspartate	9	8	8	7	8	7
Glutamate	10	11	10	9	11	10
Hydroxyproline	3	2	*	*	*	*
Serine	7	10	10	10	10	9
Glycine	19	16	9	16	15	13
Histidine	1	1	*	*	1	1
Arginine	3	4	3	2	3	3
Threonine	4	5	5	3	4	4
Alanine	10	9	7	8	9	9
Proline	5	6	4	3	4	4
Tyrosine	2++	2++	2++	4++	2++	2++
Valine	13 ‡	8	5	23 ‡	13 ‡	13 ‡
Methionine	1	2	2	*	2	2
Isoleucine	4	4	5	6	5	5
Leucine	5	7	8	5	7	6
Phenylalanine	3	3	4	2	3	3
Lysine	2	2	2	2	2	2
⊗ Percent of fraction	20	80		25	75	
⊙ Media sample, in micrograms	1	0.3				

KEY AND NOTES
S=Supernatant
P=Precipitate
N= Normalized percent composition
* Below detectable level or less than 0.5percent
+ Sample was treated with performic acid, after which methionine is converted to methionine sulfone, cysteine is converted to cysteic acid (not shown), and tyrosine is destroyed
++ Possibly tyrosine preserved in egg yolk sample, or possibly unidentified material with same elution (see fig. 2)
‡ Valine peak was overwhelmed by nonproteinaceous material eluting at approximately 9 minutes
⊗ Percent of fraction is based on total peak areas for each chromatogram
⊙ Total media sample size (approximate proteinaceous yield) based on peak areas compared with reference standards

AMINO ACID COMPOSITION OF SAMPLES FROM TWO PAINTINGS BY ANDREA MANTEGNA: *Christ Child Blessing* **(FIG. 11) AND** *Judith and Holofernes* **(FIG. 12), WITH NORMALIZED PERCENT COMPOSITION OF ANIMAL GLUE SUBTRACTED FROM THE PRECIPITATE IN COLUMN 1, AND NORMALIZED PERCENT COMPOSITION SUBTRACTED FROM THE PRECIPITATE IN COLUMN 3**
Results of the water-soluble supernatant fraction and the water-insoluble precipitate fraction are given as mole percent

COLUMN 1: Supernatant and precipitate fractions of off-white paint sample from column left of figure in fig. 11
COLUMN 2: Composition of animal glue (see table 1, column 1) normalized for hydroxyproline to equal 2 and subtracted from percent composition of precipitate of off-white sample from fig. 11

COLUMN 3: Supernatant and precipitate fractions of off-white paint sample from bottom of fig. 12
COLUMN 4: Composition of dammar (see table 1, column 10) normalized for aspartate to equal 1 and subtracted from percent composition of precipitate of off-white sample from fig. 12. The subtraction was done to remove trace contaminants from the precipitate, which constituted the majority of both the samples. The subtracted precipitates match each other as well as the composition of egg yolk (see fig. 13a–f)

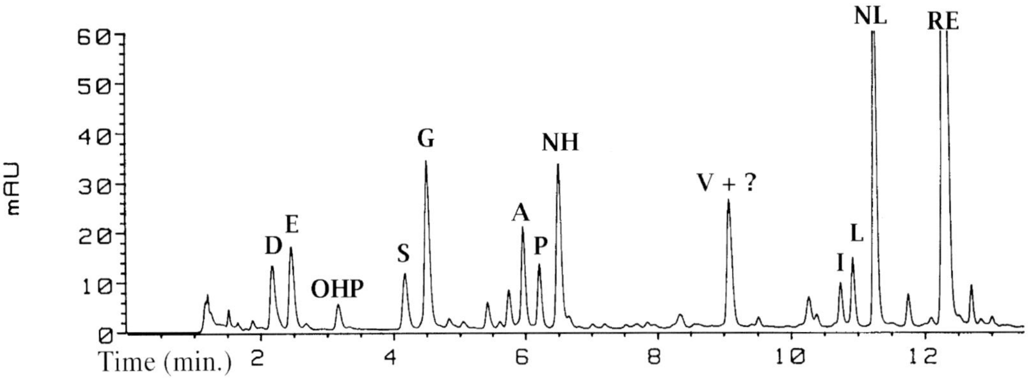

fig. 13a

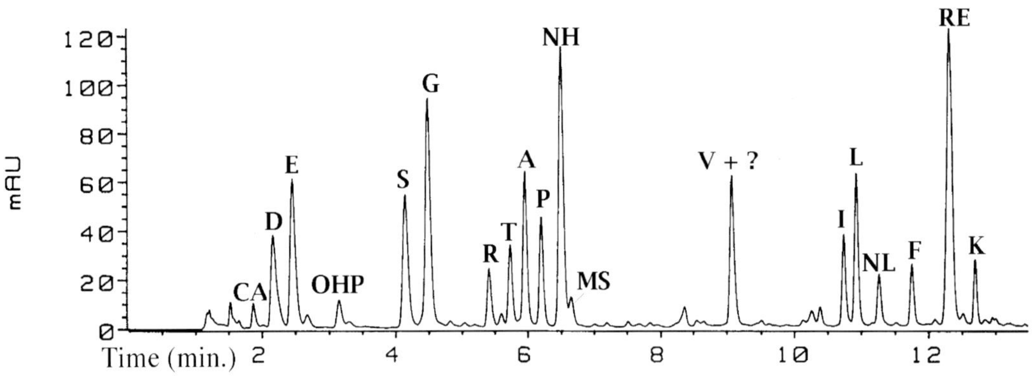

fig. 13b

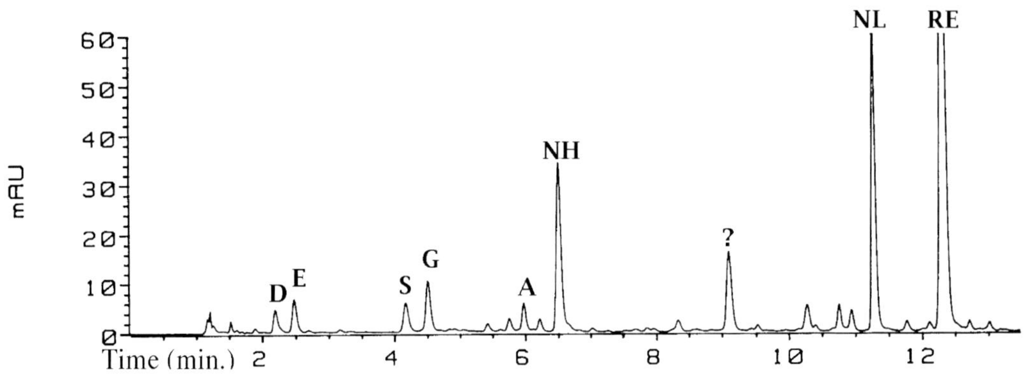

fig. 13c

13. Chromatograms of (single) samples from Mantegna, *Christ Child Blessing* and *Judith and Holofernes* (figs. 11–12, respectively). (a) Off-white paint, supernatant fraction, from *Christ Child Blessing* (fig. 11), showing small amounts of animal glue (hydroxyproline) and possibly amino acids from aged dammar (large peak at 9 minutes); (b) off-white paint, precipitate fraction, containing 80 percent of the sample, matching the composition and solubility of egg yolk. A small amount of glue (less than 20%) is also present in the precipitate, as seen by the very small hydroxyproline peak, probably coming from the hot-glue relining or from attached gesso ground. The striking difference between the supernatant and the precipitate is consistent with egg yolk paint layers on a gesso ground (see fig. 5a–b) and suggests that the egg yolk was not mixed with glue before application

(c) Off-white paint, supernatant fraction, from *Judith and Holofernes* (fig. 12), showing trace amino acids possibly from a dammar varnish (large peak at 9 minutes);

OPPOSITE PAGE

(d) off-white paint, precipitate fraction, containing 75 percent of the sample and matching the composition and solubility of egg yolk. Using the data handling system, the peak areas in supernatant were subtracted from the peak areas in precipitate of each sample to compare the composition of the paint layers without interference from the trace (less than 25 milli-absorbance units) material in the supernatant. The results from the subtraction are shown in (e) for *Christ Child Blessing* and (f) for *Judith and Holofernes*. Both "cleaned up" precipitates closely resemble the composition of egg yolk (see table 3 for percent composition of amino acids). All four samples were treated with performic acid

fig. 13d

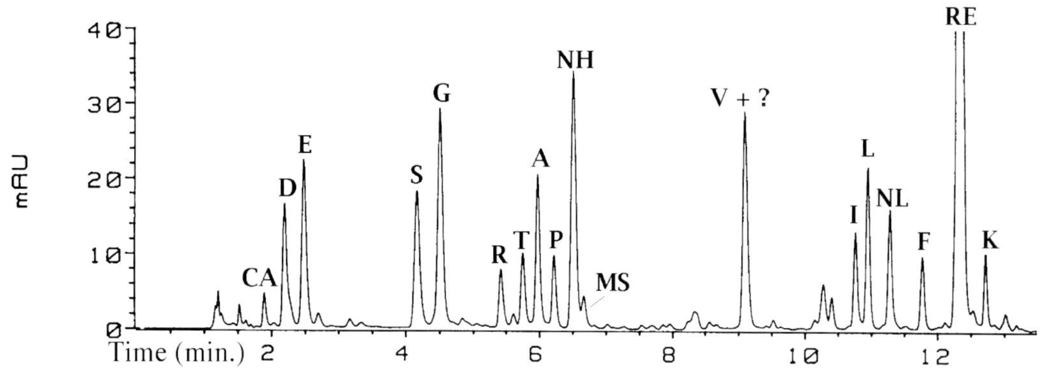

fig. 13e

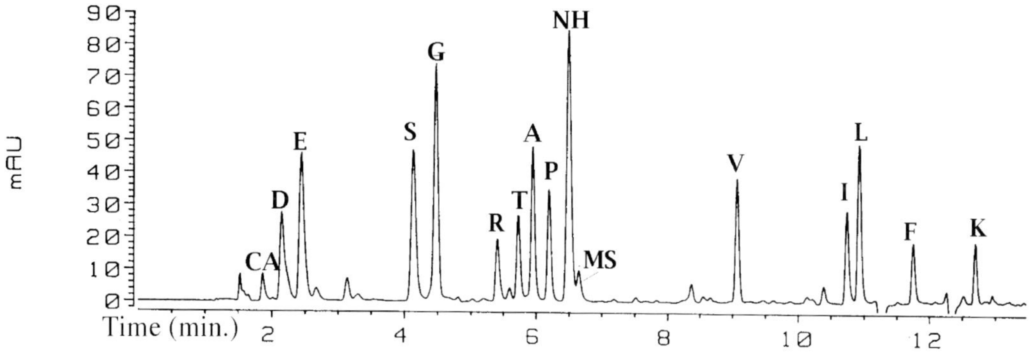

fig. 13f

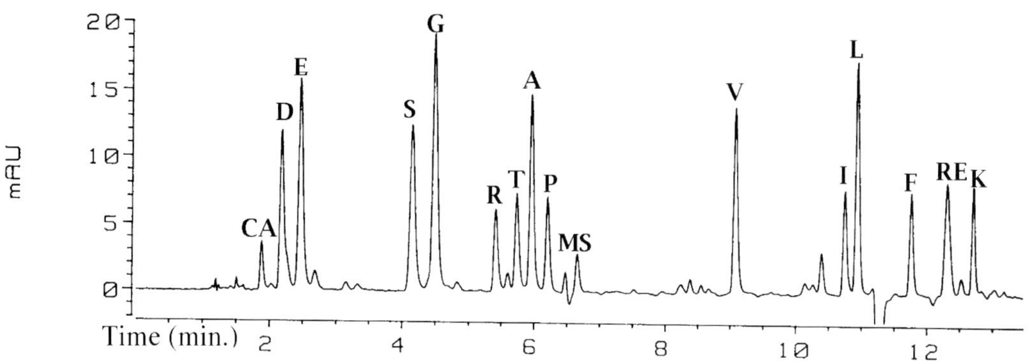

Two areas were sampled from each painting. The canvas of *Christ Child Blessing* was sampled from a brick red area to the left of the figure and from an off-white area, again left of the figure. The sampled areas on the panel of *Judith and Holofernes* were from the blue in the top right corner and the off-white in the bottom left corner.

The majority of the samples appeared in the precipitate fractions, indicating water-insoluble paint binders, and the composition of samples from both paintings suggested egg material, with traces of other materials such as glue and perhaps dammar also present (see table 3 and fig. 13a–d). The small amount of glue in the precipitate of the samples from *Christ Child Blessing* may be due to adjoining gesso ground or to a previous glue relining (table 3, column 1, and fig. 13a–b). The hardness of the canvas and the flattening of the surface suggest that the canvas was relined using glue and a hand iron.[59] The saturation

14. Cross section of paint, before staining treatment, from the white marble background, left of figure from Mantegna, *Christ Child Blessing* (fig. 11). (1) The thin gesso layer is at the bottom, covered by (2) a gray paint with (3) an off-white paint blended into it. The two white paint layers (2 and 3) seem to have been blended, in much like a wet-on-wet technique rather than a distinct layering usually found with tempera samples

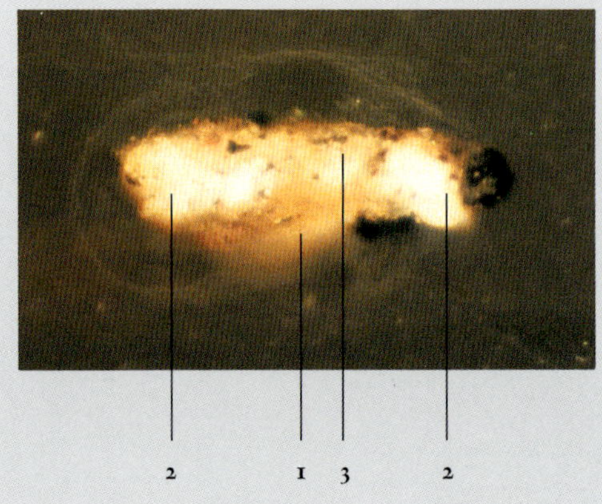

of the egg paint layer with hot glue may have made the glue less soluble than expected. A small amount of animal glue alone appeared in the supernatant, also suggesting an adjoining layer of glue or gesso, or a consolidant, but not a mixed medium. In the samples from *Judith and Holofernes*, trace amino acids associated with dammar varnish appeared in the supernatant, most likely remaining from a prior conservation treatment (table 3, column 3, and fig. 13c–d).

To clarify the composition of the sample materials, trace amino acids from glue and dammar were subtracted from the precipitate fractions, where the majority of the samples remained. Therefore normalized percent compositions of animal glue and dammar were subtracted from the off-white precipitates from *Christ Child Blessing* and *Judith and Holofernes*, respectively. The subtracted composition of animal glue (table 1, column 1) was normalized with hydroxyproline equal to 2 percent, and the subtracted composition of dammar (table 1, column 10) was normalized with aspartate equal to 1 percent. The resulting corrected compositions (table 3, columns 2 and 4) match each other closely and match the pattern for egg proteins. To further visualize the corrected yield from the two off-white samples, the peak areas in the supernatant fractions (fig. 13a and c) were subtracted from the peak areas in the precipitate fractions (fig. 13b and d) using a data-handling program. As the supernatants contained traces of animal glue and dammar, the resulting subtracted plots (fig. 13e–f) reveal very similar traces, indicating an egg binder in both paintings.

Egg appears as the major component in both sets of samples, indicating that the artist(s) used the same binder—egg yolk—to paint these two particular pieces. This finding suggests that the panel is perhaps a fairly straightforward egg tempera panel, but that the canvas painting is a most unusual example of egg tempera applied on a very thin ground layer.

The amino acid analysis results from *Christ Child Blessing* indicating egg tempera binder on canvas were corroborated by fatty acid analysis, which showed palmitate and stearate components but no azelate.[60] A cross section taken from the marble background (fig. 14) revealed a thin ground layer below the paint.[61] Furthermore, the paint layer appears to consist of two paint applications. These two paint layers were not as distinct as usually seen in egg tempera panels but undulated in what may be a wet-on-wet technique. Staining the cross section with amido black 3, which stains most readily for glues, showed heavy uptake at the bottom and to a lesser extent through the middle of the sample, a pattern consistent with the presence of a gesso ground and the possible use of a glue-based relining or consolidant (not illustrated). The two paint layers did not take up the stain. The small amount of glue that appeared associated with the egg binder in the amino acid analyses may well have been from the glue relining forced by heat through the fissures in the paint surface, as well as from the gesso ground.

Egg tempera paintings on a canvas support with a thin ground layer are somewhat unusual, but these findings are not entirely unexpected. In a section describing how to paint on cloth, Cennini advised:

> *In the same way as for anconas you should paint, step by step, on this cloth; and it is more pleasant to work on than on panel, because the cloth holds the moisture a little; and it is just as if you were working in fresco, that is, on a wall. And I will also inform you that, in painting, the colors must be laid in many, many times, far more than on panel, because the cloth has no body as the ancona has, and it does not show up well under varnishing when it is poorly laid in. Temper the colors the same as for panel. And I will not enlarge upon this any more.*[62]

As mentioned, Cennini recommended egg yolk as the principal binder for panel paintings, with animal glue specified for certain colors, and therefore these recommendations applied to paintings on cloth as well. He also suggested, "The less gesso you leave on [when preparing cloth paintings], the better it is."[63] The gesso ground is certainly very thin on this canvas painting; in fact, it is barely perceptible through the paint and heavy varnish layer but apparent in the cross section.

The cross-section sample (see fig. 14) was taken from the left side, from an area of the marble background. Mantegna may have taken advantage of the cloth's ability to hold "moisture a little . . . as if you were working in fresco," to use something like a wet-on-wet technique to get the marbleized effect. The cross section shows a white paint layer with a more heavily pigmented off-white layer blended in, resembling an oil's wet-on-wet technique more than the distinct paint layers usually associated with egg tempera paintings.[64] Working on canvas might have allowed the artist to blend certain areas, such as marble, in ways not easily achieved on panel paintings.

As a final note, Cennini seems to make a distinction between paintings on cloth, rendered primarily in egg tempera and quite possibly varnished, and hangings made of cloth, or what are now referred to as banners. The hangings are assumed to be decorations for specific festive occasions or processions, and, although they may have been prepared with great care, they were not intended as permanent works of art. In the section "Various Ways to Do Hangings," which follows the chapter on cloth paintings, Cennini instructed, "You do not have to apply gesso" but simply "lay a coat of that size with which you temper the gessos."[65] He also suggested gluing pieces of cloth on these hangings and applying washes of colors, "without varnishing afterward. And you get more done, and cheaply, and they are handsome enough at the price. Also, on hangings, you may do some foliage decorations with a brush . . . tempering with size."[66] Cennini seems to make a very clear distinction between permanent paintings on cloth and temporary hangings. Those of a permanent nature would be prepared with a ground, with the colors tempered the same as for panels, and then would be varnished. Temporary hangings were cheaply made; they lacked a gesso layer or a final varnish coat and were painted with a glue size and possibly had pieces of material glued down in place of painted colors. The thin ground layer and the egg tempera binder place *Christ Child Blessing* in the group of paintings on cloth.

Cennini mentioned varnishing the cloth paintings in almost a matter-of-fact way, whereas he specifically said that the hangings do not need to be varnished. Correspondence between Isabella d'Este and the Bolognese painter Lorenzo Costa describes Mantegna's Studiolo allegories, canvas paintings commissioned by Isabella, as "varnished after being finished" and records show that varnish was purchased for the first of the Studiolo allegories.[67] These canvas paintings also appear to be rendered with egg tempera on a gesso preparation.[68] It seems that Mantegna varnished at least some of his egg tempera on canvas paintings, and therefore *Christ Child Blessing* may have been initially varnished by the artist. It is also apparent that the heavy varnish layer(s) presently covering the painting disfigures the work.

In *Christ Child Blessing*, Mantegna demonstrates his ability to take a traditional medium and use it to explore new visual schemes such as the marbling effect achieved by blending different paint layers in what appears to be an egg tempera wet-on-wet technique.

Paul Gauguin, *Fatata te Miti*

A technical study of Gauguin's materials and technique raised questions concerning the artist's use of casein in the gesso ground preparations for his canvases. There are two

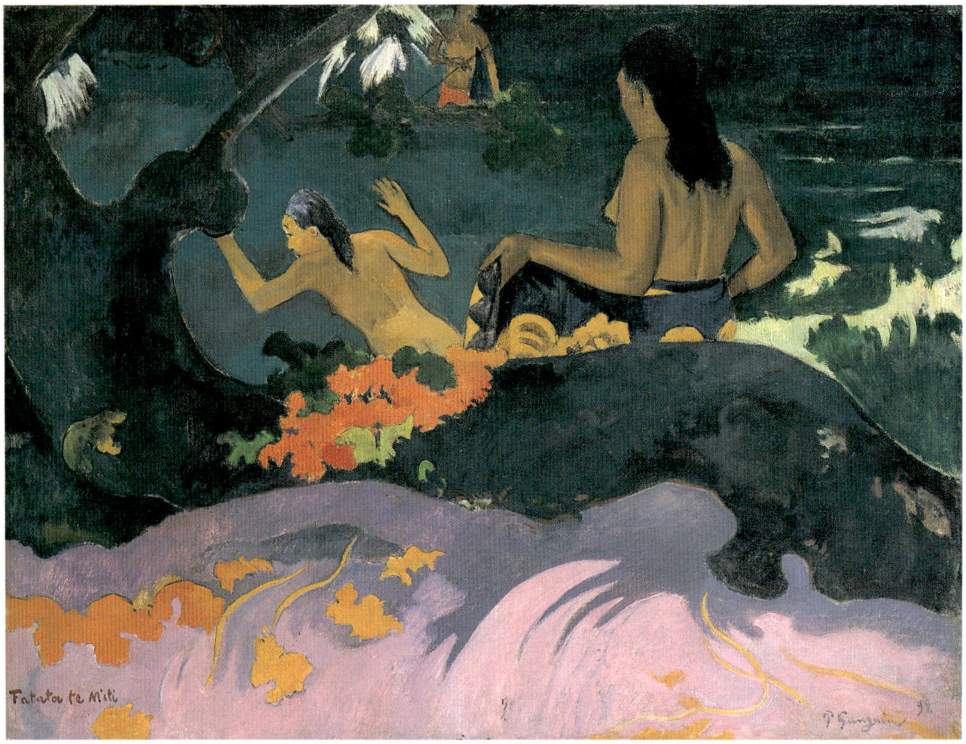

15. Paul Gauguin, *Fatata te Miti* (By the sea), 1892, oil on canvas
National Gallery of Art, Washington, Chester Dale Collection

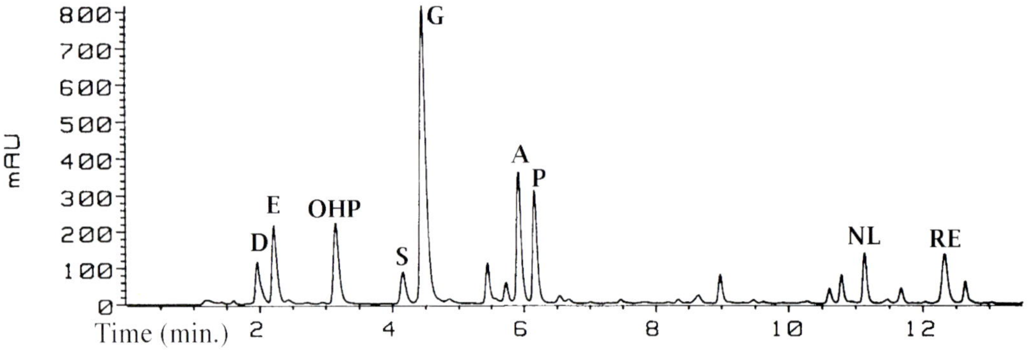

fig. 16a

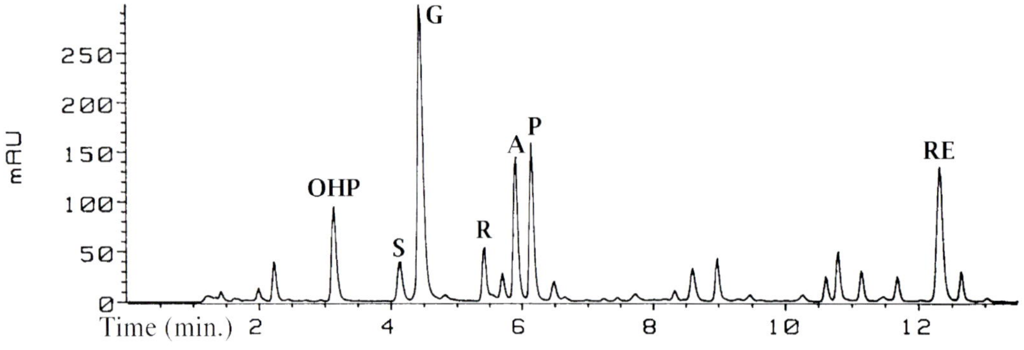

fig. 16b

16. Chromatograms of a sample of gesso ground from Gauguin, *Fatata te Miti* (fig. 15). Both (a) supernatant fraction and (b) precipitate fraction show the presence of animal glue alone, with 75 percent of the total sample appearing in the water-soluble supernatant, as expected for animal glue (compare fig. 4c–d)

17. Edward Coley Burne-Jones, *Saint Barbara*, c. 1886–1870, mixed media on paper
National Gallery of Art, Washington, Gift of William B. O'Neal

references to Gauguin's using casein: in one he was described by a contemporary as adding casein to the gesso while preparing the ground; he also requested his friends and suppliers to ship casein to him in the South Seas.[69]

Paint binders have been identified most commonly using fatty acid analysis and by staining for protein and oil.[70] Neither of these techniques would help in identifying a casein and animal glue mixture, as both components should react positive for a protein stain, negative for an oil stain. With glue having at most a 10 percent fat content[71] and casein having no significant fat content, fatty acid analysis provides little identifying information. On the other hand, amino acid analysis can be particularly useful for discerning mixtures of proteinaceous paint binders, especially if the binders have different solubilities in water and can be selectively extracted.

A sample from the ground of Gauguin's *Fatata te Miti* (fig. 15) was submitted for amino acid analysis to identify the components of the binder.[72] Both the amino acid composition of the supernatant and the precipitate (table 2, column 5, and fig. 16) matched the pattern of animal glue alone. The composition did not show any additional amino acids consistent with casein, indicating that Gauguin did not mix casein into the ground layer of *Fatata te Miti*.

Two other paintings by Gauguin (not illustrated) were investigated for the binder in the ground, *Te Pape Nave Nave* (Delectable waters), 1898,[73] and *Madame Alexandre Kohler*, 1887–1888.[74] Animal glue alone was found in the ground layer of these paintings as well. Slight variations in the compositions of the analyses of these paintings were thought to be caused by the presence of calcium in the gesso ground and trace amino acids from dammar resin varnish. As mentioned, calcium has been shown to lower the recovery of glutamate and aspartate using this amino acid technique. However, the results of the three samples analyzed show that casein was not mixed in with the animal glue into the ground layer of the three paintings analyzed here, at least not in any appreciable amount.

Edward Coley Burne-Jones, *Saint Barbara*

Thus far, proteinaceous material on wooden panel and stretched canvas has been discussed. Burne-Jones' *Saint Barbara* is painted

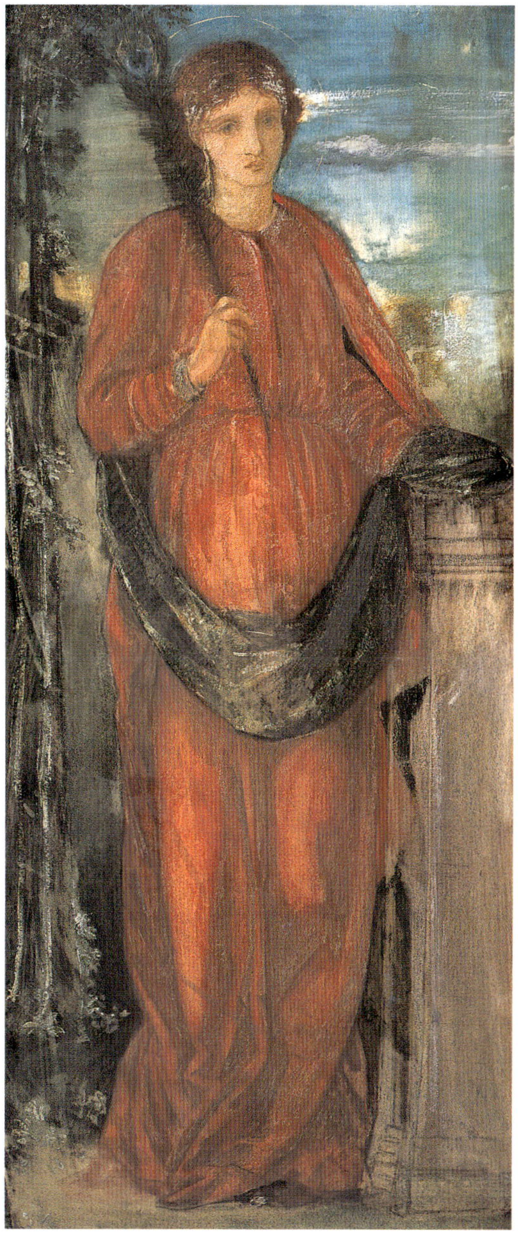

on a paper support, and the medium was thought to be anything from a drying oil to a proteinaceous binder such as casein (fig. 17). The paint binder was of great interest because Burne-Jones was associated with the pre-Raphaelites, a group of artists who worked in an industrialized age while deliberately seeking out methods they believed to have been used centuries before. Their working methods were not straightforward, as the artists had at their disposal materials unavailable to the Renaissance masters they attempted to emulate. Such an anachronistic approach could result in a confluence of old and new

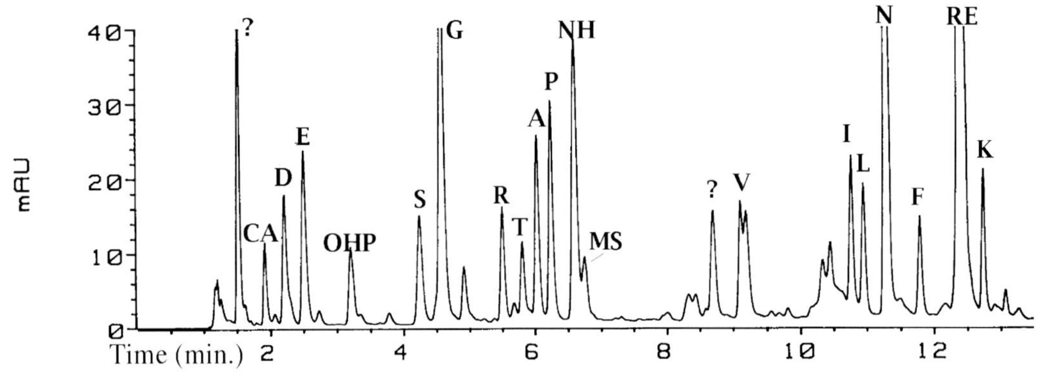

fig. 18a

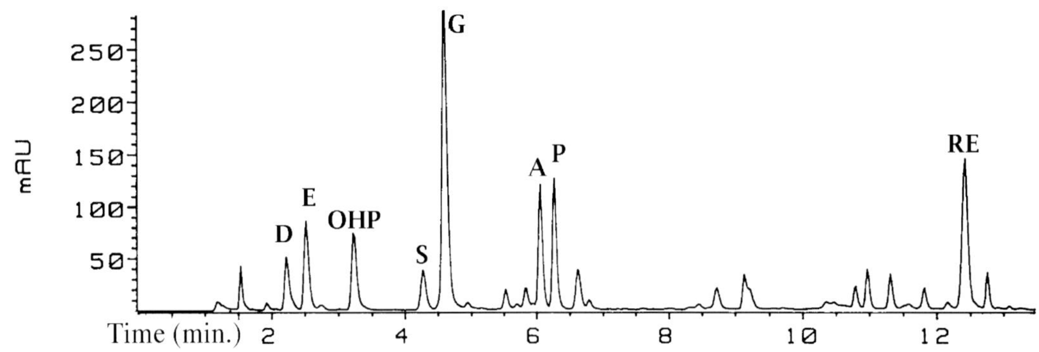

fig. 18b

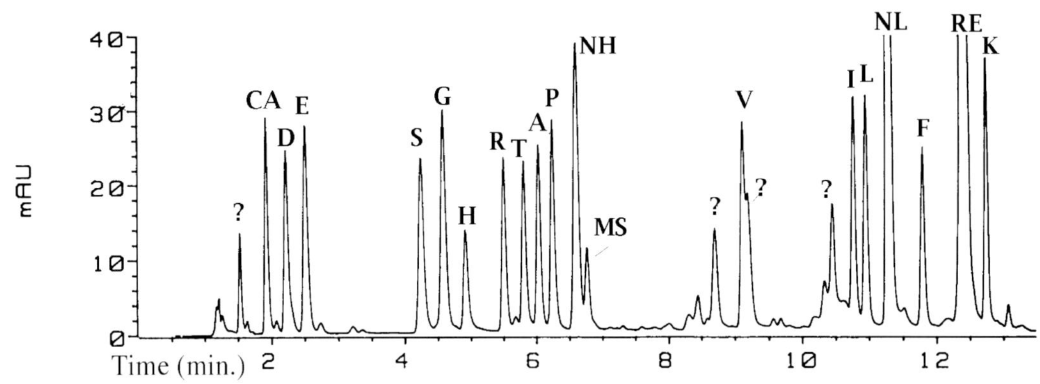

fig. 18c

18. Chromatograms of three samples from Burne-Jones, *Saint Barbara* (fig. 17). (a) Paper fibers, supernatant fractions with gelatin size and traces of egg white; (b) paper fibers, precipitate fraction, showing approximately 90 percent of the total sample, indicating that the gelatin sizing has become water insoluble

(c) Blue paint, supernatant fraction, showing the water-soluble egg white binder;

OPPOSITE PAGE
(d) blue paint, precipitate fraction, with water-insoluble gelatin (hydroxyproline in precipitate only) and traces of egg white

(e) White paint, supernatant fraction, showing trace amino acids associated with dammar resin; (f) white paint, precipitate fraction, with dammar resin and insoluble gelatin size. All samples were treated with performic acid

fig. 18d

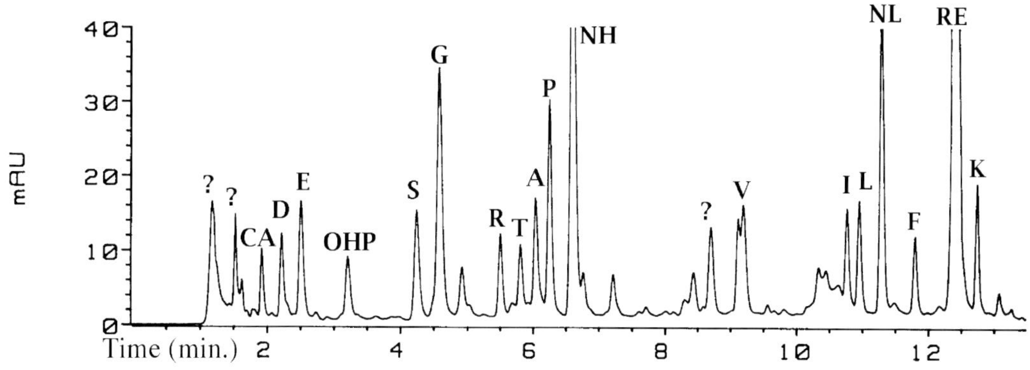

fig. 18e

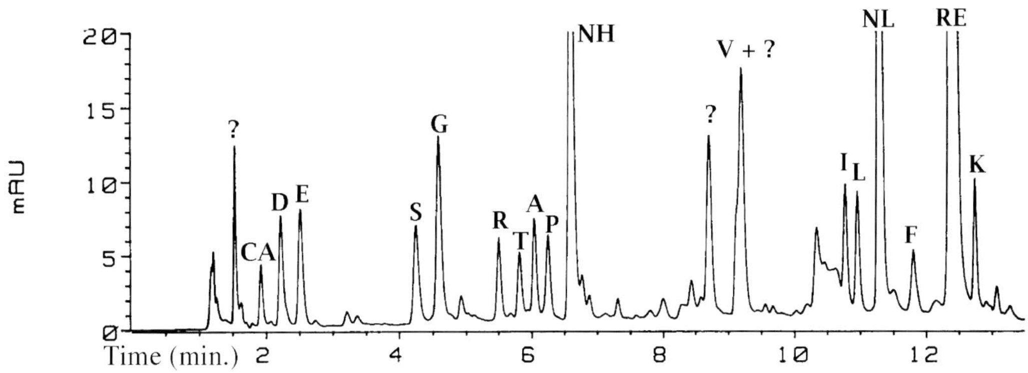

fig. 18f

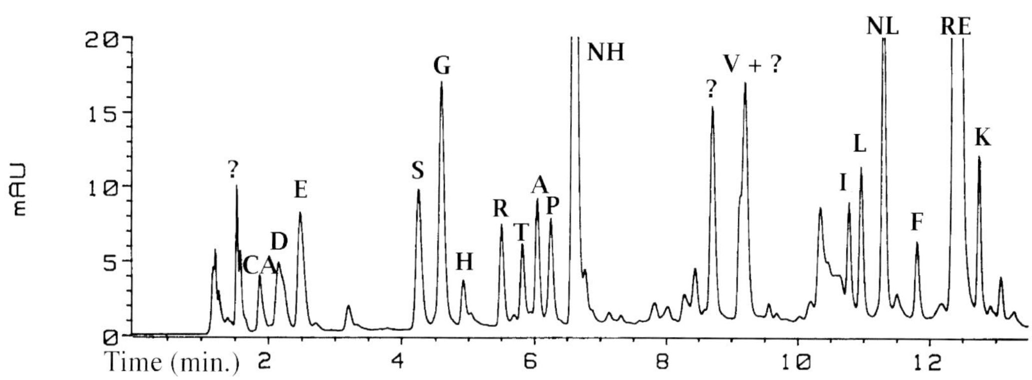

materials. Furthermore, analysis of the paint had to be approached with caution, as the paper itself might contain a proteinaceous sizing. *Saint Barbara* is a particularly complicated painting, as it seems to have begun as a design for a stained glass window and was later reworked and given as a wedding gift to Burne-Jones' sister-in-law.[75]

To determine the nature of the binding medium, samples from *Saint Barbara* were submitted for amino acid and fatty acid analysis. Cross-sectional analysis is usually impossible for paintings on paper and is not recommended.

To investigate the nature of the paper sizing, a sample of paper fibers was analyzed separately for amino acids, followed by analysis of five painted areas of a different color and appearance.[76] The areas sampled were the blue from the sky, red from the top of the robe, white from the sky (including some blue paint and paper fibers), a red glaze from the bottom of the robe, and a "waxy" red line from the right of the subject's neck.

The results were somewhat unexpected, as the solubility of the materials was the opposite of what is usually seen in tempera panels. Animal glue appeared in the insoluble precipitate (table 2, columns 6–8, and fig. 18b, d, and f) of all six samples rather than in the supernatant, and egg protein appeared in the soluble supernatant (table 2, column 7, and fig. 18c) of two paint samples.[77]

A closer look at the sample location helps clarify the unusual solubility of the materials. The paper fibers from an unpainted edge showed a large amount of animal glue in the precipitate (table 2, column 6, and fig. 18a). Its presence indicates that the paper contained a gelatin size. The gelatin may have become insoluble by binding with reactive sites on the paper fibers. A high number of aldehyde groups were found to develop in aged paper.[78] In the paper industry, formaldehyde is often added to gelatin size to improve the water resistance of paper,[79] and the naturally occurring aldehydes and/or other reactive groups may have had a similar effect on the gelatin size of this paper support. Although the chemical interactions have not been established, the gelatin sizing remained in the water-insoluble precipitate after the 1h extraction method instead of in the soluble supernatant fraction, as usually seen with glue-containing samples.

Amino acid analysis results of the blue from the sky (table 2, column 7, and fig. 18c–d) and red from the top of the robe showed egg proteins in the supernatant and glue in the precipitate. Egg yolk is water insoluble and appears in the precipitate; therefore, the appearance of egg material in the supernatant (fig. 18c) was very unusual. Gas chromatographic analysis of several samples, including the blue area in the sky, showed traces of fatty acids consistent with cow or sheep hide glues, but not with the palmitate-to-stearate ratio of egg yolk.[80] Therefore it appears that these areas were applied with egg white, not egg yolk, which remained water soluble on the surface of the gelatin sizing.

The white sample from the sky showed the low amino acid pattern associated with dammar varnish (table 2, column 8, and fig. 18e–f). This white area was also shown, by oil and resin analysis, to contain triterpenoids.[81] The remaining two samples from the red glazed area and the red line near the neck showed the gelatin size in the precipitate and an additional nonproteinaceous material that could not be identified by oil or resin analysis.[82]

These results indicate that Burne-Jones used egg white, perhaps dammar, and an unidentified paint as binders on the same painting. The complicated layering of materials may be in part due to the artist's reworking of a stained glass design into a finished painting. Amino acid analysis using the 1h extraction method identified the paper sizing and the presence of egg white paint while suggesting the presence of dammar and a nonproteinaceous binder, thus giving insight into this mixed media piece in a single analytical technique.

Henri de Toulouse-Lautrec, *Maxime Dethomas and Lady with a Dog*

The cardboard support from Toulouse-Lautrec's *l'essence* paintings, *Maxime Dethomas* (fig. 19) and *Lady with a Dog* (fig. 20), was analyzed as part of a comprehensive study of the artist's techniques.[83] The cardboard used was believed to have been an inexpensive material, mass produced, and sized with casein. It is unlikely that the boards were intended as artists' supports, since the surfaces are often marred by a loose, random burlap imprint, presumably made under pressure during the manufacture of the board. Far from avoiding these textured areas, Toulouse-Lautrec often centered his figures on them.[84] The substantial number of Toulouse-Lautrec's works on cardboard and the way in which the support is allowed to show through the paint strokes attest to his preference for this support for aesthetic, not economic, reasons.

The samples were collected by the conservator from the pieces of the cardboard support that had accumulated inside the frame of the two paintings. The samples (c. 1 x 2 mm) generally consisted of one lamina each. The analytical results from samples from *Maxime Dethomas* and *Lady with a Dog* were nearly identical; therefore, the discussion of their results will be combined.

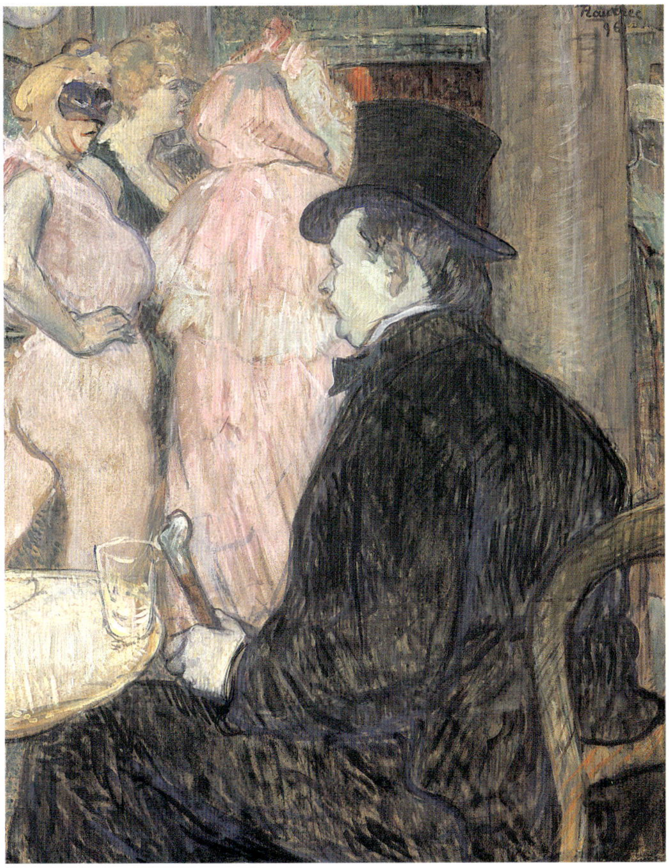

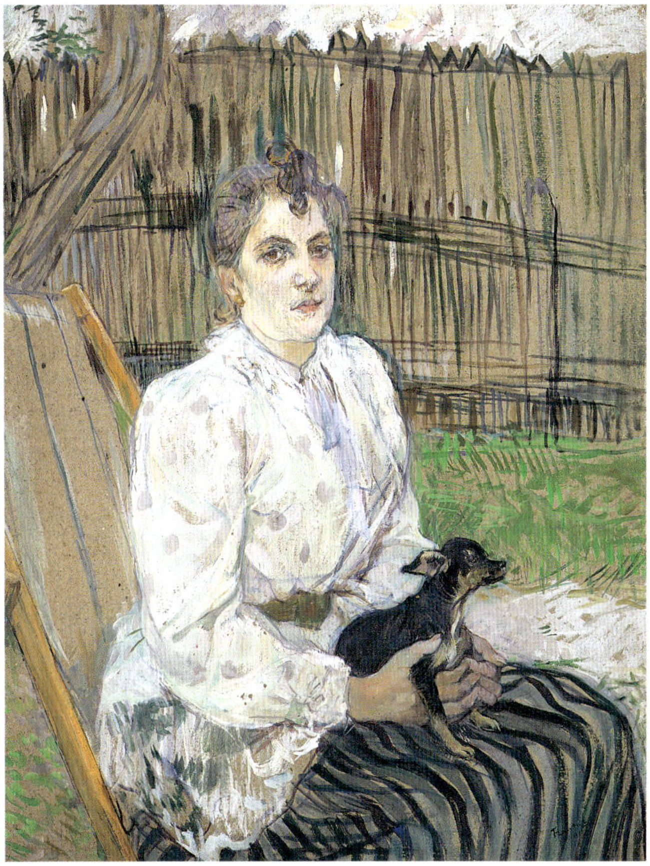

19. Henri de Toulouse-Lautrec, *Maxime Dethomas*, 1896, oil on cardboard
National Gallery of Art, Washington, Chester Dale Collection

20. Henri de Toulouse-Lautrec, *Lady with a Dog*, 1891, oil on cardboard
National Gallery of Art, Washington, Gift of the W. Averell Harriman Foundation in memory of Marie N. Harriman

Fiber analysis showed that the cardboard contained an assortment of hardwood, softwood, and grass fibers, a combination consistent with a low-grade paper.[85] The appearance and uniform thickness suggested that the cardboard had been mass-produced.

Results from resin and fatty acid analysis indicated the presence of a nondrying oil (palmitate and stearate, with no azelate).[86] The palmitate-to-stearate ratio matched that of some animal glues rather than that of egg yolk. These analyses did not detect any evidence of rosin or resinous materials present in the samples.

The amino acid composition showed the presence of animal glue in the water-soluble fraction (table 2, column 9, and fig. 21a) and an additional water-insoluble proteinaceous material in the precipitate (table 2, column 9, and fig. 21b). A more elaborate extraction procedure revealed that indeed two proteinaceous materials are present, a soluble animal glue and a less-soluble unknown.[87] The unknown material is not readily discernible, but certain anomalies such as the high alanine and low proline peaks rule out casein. Rather, the composition resembled that of an albuminous protein such as ovalbumin, found in eggs. It is unlikely, however, that French papermakers would have used eggs, especially egg whites, to mass-produce inexpensive cardboard. The close match between the results from the two supports for both fatty acid and amino acid analyses suggest a uniform manufacturing process and not inadvertent additions. Furthermore, the somewhat reduced amounts of the three basic amino acids—histidine, arginine, and lysine—indicate a harsh processing treatment.

In searching for other insoluble globular proteins, albumin or blood glue was considered as a possible size used in the cardboard. The published composition of bovine serum albumin[88] is indeed quite similar to egg material. Table 1, column 9, shows the composition of blood albumin; it should be noted that albumin is the largest component but not the only component in blood. An adhesive specialist has indicated that a high iron content in the cardboard would suggest the presence of blood glue, as it is the only iron-containing proteinaceous glue and the resin glues are unlikely to have a high iron content.[89] Analysis of an unpainted surface of the cardboard using x-ray fluorescence showed that iron was present in significant amounts.[90] Could blood glue have been used as a cardboard adhesive in late nineteenth-century France?

"It is known that all sorts of things can be glued with blood or albuminous glues": so begins a 1936 French patent for a blood glue containing urea and formaldehyde.[91] Blood glue has been used since prehistoric times. The Chinese, Aztecs, and Russian slaves were all alleged to have mixed blood in with cement for a more permanent mortar, and the United States asphalt industry used blood to stabilize an asphalt-water emulsion in the 1950s and 1960s.[92] Blood glue has been used throughout European history "shrouded in the secrecy of proprietary knowledge," was especially used in the Baltic countries, and has often been mixed with other glues to enhance the tack, flow, or general consistency of the glue.[93] *The Techno-Chemical Receipt Book* published in 1887 describes the use of blood as a sizing for cloth, and as late as 1961 it was still used as a pigment binder for printing on cotton.[94]

Concurrent with the development of the plywood industry at the turn of the century, there was an increased interest in "new" forms of adhesives in Europe and the United States. Demonstrations of plywood furniture in nineteenth-century trade exhibitions promoted the manufacture of plywood from 1850 onward.[95] Blood glue was a choice adhesive for plywood as it has been described as the most water-resistant adhesive available prior to the synthetic resins.[96] From 1905 to 1981 at least fifty patents dealing with blood glues

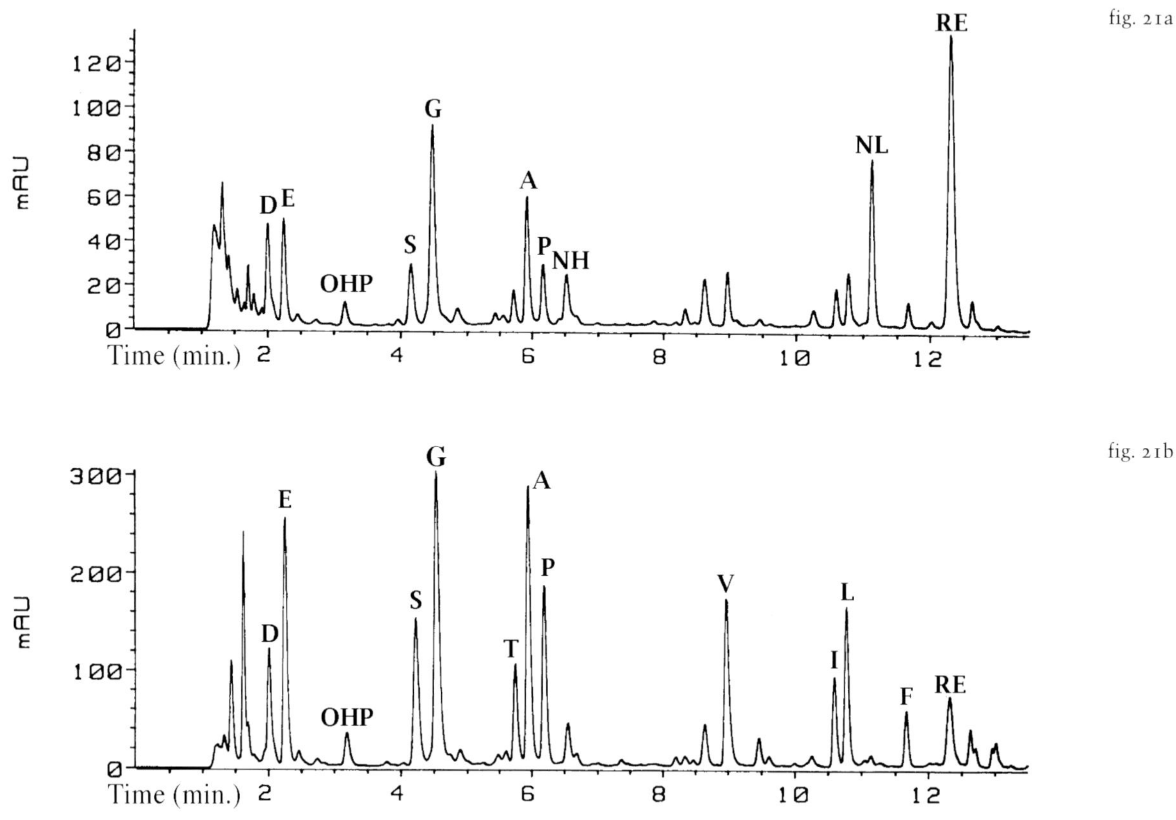

21. Chromatograms from a sample of cardboard support from Toulouse-Lautrec, *Maxime Dethomas* (fig. 19). (a) Supernatant fraction showing the animal glue size component, and (b) precipitate fraction, showing what may be the albuminous component from the blood glue sizing of the cardboard. More elaborate extraction procedures confirmed the presence of two main proteinaceous components, consisting of approximately one-third glue and two-thirds unknown, possibly serum albumin

were issued in Germany, France, Great Britain, and the United States.[97] The cost of blood glues, especially the fresh material—that is, not dried—and their water-resistant properties, enhanced with hot-plate processes more common in Europe, made it an ideal adhesive for plywood and cardboard.[98] Because of the demand for a water-resistant plywood for making aircraft, the use of blood glue rose sharply during World War I.[99]

At one time proteins, including collagen or animal glue, casein, and soluble dried blood, were the most important adhesive material used in the paper industry.[100] A British patent issued in 1923 describes a "cement comprising a mixture of blood albumin and gelatine, or animal glue, combined with formaldehyde" for use in the manufacture of plyboard, cardboard, paper, and other absorbent material in sheets.[101] Several other patents specifically mention paper, paperboard, pressboard, cardboard, and cellulose products in general as suitable applications for their albumin glues.[102] A French patent of 1966 describes the use of fresh blood combined with sulfuric acid in the preparation of an adhesive for paper and cardboard.[103] As late as 1979 a United States patent was issued for a blood glue with organic acid additives, suitable for the manufacture of cellulose products such as chipboards and paper bags.[104]

Based on the sample analyses and the established use of blood glue for manufacturing cellulose products, it is likely that a mixture of blood and animal glue was used as the size in the cardboard support of the two paintings by Toulouse-Lautrec. Confirmation of blood by the traditional forensic methods is not reliable because these methods depend on the identification of the heme group, which may have been destroyed by heat and the strong alkaline reagents often used to prepare the glue and which would be further reduced over time.[105] Strong reagents may also have been involved in reducing the yields of the basic amino acids as well, but the mechanism for this is not fully understood.

The unusual combination of blood albumin and animal glues found in the Toulouse-Lautrec paintings, *Maxime Dethomas* and *Lady with a Dog*, may possibly be used to authenticate similar cardboard supports made during the same period in France.

Conclusion

Amino acid analysis offers significant information that cannot be obtained by fatty acid analysis and cross-sectional staining. Before the use of amino acid analysis, which this research introduced to the conservation field as a routine analytical tool, identification of proteinaceous binders could not be done with any degree of certainty. Different proteinaceous media, whether distemper and tempera used side by side or in layers, mixtures of casein and animal glue in ground preparations, or paper sizing with casein or blood albumin, cannot be detected by fatty acid analysis, while confounding factors such as the age of the material and associated pigments can hamper accurate cross-sectional staining methods. Amino acid analysis of proteinaceous artists' materials provides insight into the nature of the binding medium, the intent of the artist, and the original appearance of the paint surface, while it helps explain the susceptibility of certain materials to flaking and peeling.[106]

Combining amino acid analysis with the 1h extraction procedure allows for a clearer identification of the material by separating the proteins based on water solubility and removing interfering pigments and other insoluble contaminants to the precipitate. The water extraction also helps to determine if mixtures of materials are coming from adjoining layers or from a single mixed medium layer. More explicit results can indicate the nonproteinaceous components as well.

Media analysis of Gaddi's *Madonna Enthroned with Saints and Angels* indicated that, like Cosimo Tura, the artist's choice of binder was dependent on the pigment being used. Gaddi also seems to have employed an unusual sgraffito painting technique in the red brocade, which consisted of a layer of distemper covered by an egg tempera glaze. Analysis of Mantegna's *Christ Child Blessing* revealed a somewhat unexpected egg tempera binder on canvas support, although other Mantegna egg tempera canvases have been reported. Burne-Jones' painting on paper, *Saint Barbara*, presented a challenge during the interpretation of the results. The piece appears to have been rendered with a variety of binders, including a possible gelatin size that interacted with the paper support fibers and remained water insoluble; a water-soluble egg white binder; dammar resin, revealed by its trace amino acids; and an unidentified binder, perhaps a gum. In fact, dammar seemed to be present in a number of the samples discussed here, and familiarity with the trace amino acids pattern has helped avoid possible misinterpretation of the results.

Identification of mixtures of proteinaceous binders is well suited to amino acid analysis. Results from analyses of three of Gauguin's canvases showed that he probably did not use a casein and animal glue mixture in the ground layers, at least not on a consistent basis. Investigation of the cardboard support used by Toulouse-Lautrec led to the identification of an animal glue and blood albumin mixture as the possible adhesive in the cardboard. These results emphasize the importance of interpreting analytical results within a proper context: although the insoluble component from the cardboard could have been interpreted as egg white, such an interpretation seemed inappropriate within the context of the material. In addition, the antiquated use of blood as an adhesive underscores the possibility that similar technologies have passed out of our frame of reference, even within one generation.

Several aspects of the amino acid technique should be further developed. The analytical system described here is sensitive enough to recognize trace amino acid patterns in nonproteinaceous materials (c. 200 ng.). Some work has been done in characterizing trace amino acids found in starch pastes, gums, and resins, but further research must be carried out before these trace patterns can be used to characterize samples. In addition, a systematic study of the effects of pigments on the analytical results should be undertaken. Extraction procedures using other solvents might be introduced to separate nonproteinaceous materials for further analysis. Research into a system that analyzes amino acids and fatty acids during a single chromatographic analysis is under way. Such a method is needed to characterize emulsions such as egg yolk and egg-and-oil mixtures and would reduce the number of samples required of an object under analysis.

The flexibility of an HPLC system is well suited for the variety of materials encountered in art conservation and places a large number of analytical techniques at the disposal of the conservation scientist. The HPLC installed at the National Gallery of Art is routinely used for amino acid analyses, but in less than a day the instrument has been adapted for organic dye and fatty acid analyses by changing the column and the solvent system.[107]

The results from amino acid analyses described here have augmented our understanding of tempera painting and other artistic techniques. Research into additional methodologies using the HPLC system will further broaden the analytical means available to the conservation scientist.

NOTES

I would like to thank the following National Gallery of Art staff for their helpful discussions and for submitting samples: in the scientific research department, E. René de la Rie, Barbara Berrie, Michael Palmer, Lisha Glinsman, and especially Suzanne Quillen Lomax for the GC–MS work; in the painting conservation department, Sarah Fisher, Carol Christensen, Ann Hoenigswald, Cathy Metzger, Michael Swicklik, and Steve Wilcox, frame conservator, in addition to former staff members Mary Bustin, Culpeper Fellow, Paula DeCristofaro, and Teresa Longyear; Shelley Fletcher and Pia DeSantis Pell from the department of paper conservation; Janice Gruver, conservation division editor; and Richard Carafelli, photographer, imaging and visual services department. I would also like to thank Stacy C. Marsella, Rutgers University, and Diane Wolfthal, Manhattanville College. Without their assistance, this work would not have been possible.

The author's fellowship at the National Gallery of Art, Washington, was sponsored by the Andrew W. Mellon Foundation.

1. Susana M. Halpine, "A New Amino Acid Analysis System for Characterizing Small Samples: Identification of Egg Tempera and Distemper in a Painting by Cosimo Tura," *Studies in Conservation* 37 (February 1992), 22–38.

2. Cennino Cennini, *The Craftsman's Handbook, "Il Libro dell'Arte"* (1437), trans. Daniel Thompson, Jr. (New York, 1966), 32–33, 36, 68.

3. Mauro Mateini, Laboratorio Scientifico, Opificio della Pietre Dure, personal conversation, Florence, May 1990.

4. Robert R. Jones, "HPLC Keeps Growing among the Separation Scientists: Among Chemists High-Performance Liquid Chromatography Is the Clear Leader in Chemical Analysis Methods," *R & D Magazine* 33, no. 1 (January 1991), 54–57.

5. Halpine 1992, 22–23.

6. Susana M. Halpine, report on ancient Egyptian objects from the Museum of Fine Arts, Boston, 27 June

1991, National Gallery of Art, Washington, scientific research department. Hereafter, in-house National Gallery of Art scientific research department reports will be noted by scientist's name (and title when first cited), subject, date, and/or file number.

7. Susana M. Halpine, report on Benozzo Gozzoli, *Crucifixion*, and other samples from the Uffizi Gallery, Florence, 12 February 1991, NGA; Diane Wolfthal, *The Beginning of Netherlandish Canvas Painting: 1400–1530* (Cambridge, 1989), 12, 26; Andrea Rothe, "Joining the Objective Facts of Science to the Subjective Realm of Aesthetic Experience," *J. Paul Getty Trust Bulletin* 4, no. 1 (1989), 3.

8. Susana M. Halpine, report on New Mexican *retablo, The Crucifixion*, from the Smithsonian Collection, Washington, 10 June 1991, NGA.

9. Cennini 1966, 32–33, 36, and 68; Halpine 1992, 28–38; Susana M. Halpine, report on Agnolo Gaddi, *Madonna Enthroned with Saints and Angels*, 23 January 1990, NGA.

10. Carol Mancusi-Ungaro, "The Rothko Chapel: Treatment of the Black-Form Triptychs," *Cleaning, Retouching and Coatings: IIC Preprints of the Contributions to the Brussels Congress*, International Institute for Conservation of Historic and Artistic Works (London, 1990), 134–137.

11. For information on glue used in carpentry, see Irving Skeist, *Handbook of Adhesives*, 2d ed. (New York, 1977), 145. For glue used to seal canvas, see Cennini 1966, 103; Wolfthal 1989, 24; Ralph Mayer, *The Artist's Handbook*, 4th rev. ed. (New York, 1981), 241, 243–345. For glue used to seal paper, see Skeist 1977, 148. For glue used to seal stone, see Cennini 1966, 118. For glue used to seal wooden panels before painting, see Pierre Garcia, *Le métier du peintre* (Paris, 1990), 133–134. For glue in gesso grounds on paintings, see Cennini 1966, 67. For glue as varnish, see J. Menei, "La restauration des fresques de Michel-Ange à la Chapelle Sixtine," *Science et Technologie de la Conservation et de la Restauration* 1 (April 1988), 91; Cennini 1966, 68. For glue used in frames, see Deborah Bigelow, "Gold Leaf on Furniture: Its History, Application, and Conservation" (Advanced diploma diss., London College of Furniture, September 1982), 35–36, 61–67, 77. Glue used in conservation was described by Sarah Fisher, head of painting conservation department, National Gallery of Art, Washington, personal conversation, 1991.

12. Pliny the Elder, *Natural History* 28, 71, trans. W.H.S. Jones (Cambridge, Mass., 1963), 159–160: "The finest glue is made from the ears and genitals of bulls. . . . The most reliable glue comes from Rhodes, which is used by painters and physicians." Additional references include 32. 73, p. 509: "It [sturgeon skin or belly] ought, however, to be cut up and soaked in water or vinegar for a night and a day, and then to be pounded by sea pebbles, to make it melt more easily"; and 32. 27, p. 515: "Fish glue removes wrinkles and fills out the skin; prepared by boiling down in water for four hours and then kneading until liquid like honey."

Theophilus, *On Divers Arts* (c. 1122), trans. John G. Hawthorne and Cyril Stanley Smith (New York, 1979), 36–38, describes the preparation of sturgeon glue for applying gold and silver in books and other surfaces and mentions that calf vellum, eel bladder, and the head bones of dried pike could be substituted for the sturgeon bladder.

Cennini 1966, 66–68, discusses specific uses of glues and provides several glue recipes. Fish glue was recommended for mending "fine paper, wooden, or bone objects"; glue made from "clippings of goats' muzzles, feet, sinew, and many clippings of skins" was suggested for wood and tempering gesso; and glues made of goat and sheep parchment were considered the best size for tempering blue pigments for panel or wall painting.

13. Henri Louis Duhamel du Monceau, *L'art de faire différentes sortes de colle* (Paris, 1771), 4, 14–15, 19–20. See also Bengt Skans, "What Do the Altarpieces of O. Vram Tell Us?" (Ph.D. diss., University of Lund, Sweden, 1991), 162.

14. Marion F. Mecklenburg, "Some Mechanical and Physical Properties of Gilding Gesso," in *Gilded Wood: Conservation and History*, ed. Deborah Bieglow, Elisabeth Cornu, Gregory J. Landrey, and Cornelius van Horne (Madison, Conn., 1991), 163–170; Mecklenburg, "The Effects of Atmospheric Moisture on the Mechanical Properties of Collagen under Equilibrium Conditions," *American Institute for Conservation Preprints*, 16th Annual Meeting (Washington, 1988), 231–244.

15. Walter Strauss, in Albrecht Dürer, *The Painter's Manual*, trans. Walter Strauss (New York, 1977), 7: "The skilled artisans and craftsmen of the Middle Ages guarded their knowledge of processes and methods with great jealousy."

16. Halpine 1992, 25.

17. Brian A. Bidlingmeyer, Steven A. Cohen, and Thomas L. Tarvin, "Rapid Analysis of Amino Acids Using Pre-Column Derivatization," *Journal of Chromatography* 336 (1984), 93–104; Steven A. Cohen, Brian A. Bidlingmeyer, and Thomas L. Tarvin, "PITC Derivatives in Amino Acid Analysis," *Nature* 320 (1986), 769–770; Steven A. Cohen, Michael Meys, and Thomas L. Tarvin, *The Pico-Tag Method: A Manual of Advanced Techniques for Amino Acid Analysis*, Millipore Corporation, Waters Chromatography Division (Milford, Mass., 1989), 2–3, 11–12.

18. Halpine 1992, 25–28.

19. G. C. Barrett, *Chemistry and Biochemistry of the Amino Acids* (London, 1985), 188.

20. For discussion of fats in animal glues, see Bengt Skans and Peter Michelsen, "Die Bedeutung von Fett in Tierleim für Malzwecke," *Maltechnik-Restauro* 92 (1986), 63–71; John Mills, "The Paint Medium," in Christopher Brown, David Bomford, Joyce Plesters, and John Mills, "Samuel van Hoogstraten: Perspective and Painting," *National Gallery Technical Bulletin* (London) 11 (1987), 85.

21. Suzanne Quillen Lomax, organic chemist, report on fatty acid analysis of animal glues using gas chromatography–mass spectrometry (GC–MS), 2 April 1992, NGA.

22. John S. Mills and Raymond White, *The Organic Chemistry of Museum Objects* (London, 1987), 76.

23. Halpine 1992, 25, 30, 32–33.

24. Sheldon Keck and Theodore Peters, "Identification of Protein-Containing Media by Quantitative Amino Acid Analysis," *Studies in Conservation* 14 (1969), 75–82.

25. Albumin makes up approximately 60% of the proteinaceous component in blood. *Kirk-Othmer Encyclopedia of Chemical Technology* (New York, 1978), 4:26.

26. Mills and White 1987, 142; Suzanne Quillen Lomax, report on fatty acid analysis of several egg yolk films using gas chromatography, 17 July 1992, NGA.

27. Barrett 1985, 188.

28. For analysis of amino sugars, see Cohen, Meys, and Tarvin 1989, 45–47. For analysis of phosphoserine, see Lola Cohen-Solal, Jane B. Lian, Dora Kossiva, and Melvin L. Glimcher, "Identification of Organic Phosphorus Covalently Bound to Collagen and Non-Collagenous Proteins of Chicken Bone Matrix," *Biochemical Journal* 177 (1979), 85. For chromatographic conditions for separating phosphoserine, see Cohen, Meys, and Tarvin 1989, 54.

29. William Stadelman and Owen J. Cotterill, *Egg Science and Technology*, 3d ed. (Westport, Conn., 1986), 108–117; Andrew D. Nisbet et al., "The Complete Amino Acid Sequence of Hen Ovalbumin," *European Journal of Biochemistry* 115 (1981), 339.

30. The percent composition of tryptophan and cysteine, respectively, in egg yolk is 1.5 and 1.8%; in egg white, 1.5 and 2.4%; in casein, approximately 1.2 and 0.7%, as estimated from composition of parmesan cheese, calculated from values published in *Composition of Foods, Dairy and Egg Products*, U.S. Department of Agriculture, Agricultural Handbook 8-1 (Washington, 1976), items 01–125, 01–124, 01–033. The percent compositions of other amino acids are listed in table 1.

31. Hydrolysis method for tryptophan recovery: add 10% thioglycolic acid to 6N (normal) hydrochloric acid, using vapor phase hydrolysis at 165°C for 45–55 minutes, described in Yoshiko Yokote, Kunio Murayama Arai, and Kiso Akahane, "Recovery of Tryptophan from 25-Minute Acid Hydrolysis of Protein," *Analytical Biochemistry* 152 (1986), 245–249; Hiroshi Matsubara and Richard M. Sasaki, "High Recovery of Tryptophan from Acid Hydrolysates of Proteins," *Biochemical and Biophysical Research Communications* 35 (1969), 175–181. For chromatographic conditions for separating tryptophan, see Cohen, Meys, and Tarvin 1989, 50.

32. Thomas Wheat, Waters Associates of Chicago, Millipore Corporation, personal conversation, 1990; Barrett 1985, 394–395.

33. For quantitative analysis of cysteine and methionine using the performic acid treatment, see Cohen, Meys, and Tarvin 1989, 8–9, 30.

34. Casein was purchased from Sigma Chemical Co. Seven-year-old egg yolk, egg white, and whole egg were painted-out on glass panels and exposed to sunlight through a southern facing window. Egg material was supplied by E. René de la Rie, prepared at the Metropolitan Museum of Art, New York.

One-month-old samples of egg yolk, glair, and whole egg were painted-out on glass panels by the author. Glair was prepared as described in Cennini 1966, 100.

35. Barrett 1985, 376–396.

36. Halpine 1992, 25, 30, 32–33.

37. The extraction was tested with samples from an egg tempera panel prepared by Barbara Berrie, conservation scientist, National Gallery of Art, Washington. The ground was prepared with rabbit skin glue and calcium carbonate. The pigments were mixed with egg yolk. A final coat of Krylon was applied as a protective, nonproteinaceous varnish.

38. It may be possible to combine the internal standard, norleucine, in the 30 microliters of HPLC-grade water added during the extraction procedure. This would save time by eliminating a step, but it has not yet been tested.

39. Several types of nonproteinaceous materials were analyzed for amino acids: gum arabic and tragacanth, supplied by Eastman Kodak and Kremer-Pigmente, a purveyor of art materials for conservation; wheat paste (Aytex) from General Mills and *zin shofu*, or Japanese precipitated wheat paste, supplied by Conservation Materials; dammar (#600 and #6000), mastic (#601), Chios mastic (#6005), and rosin (#613) from Kremer-Pigmente. E. René de la Rie provided a film of dammar resin with 2% Tinuvin 292 added as an ultraviolet absorber, painted-out on a glass plate and aged 1,400 hours in a fadeometer at the Metropolitan Museum of Art, New York. Michael Palmer, conservation scientist, provided a film of dammar resin without additives, painted-out on a glass plate and aged 1,200 hours in a weatherometer. Suzanne Quillen Lomax supplied samples of drying oils painted-out on glass plates, with 5 years of natural aging, including cold-pressed, boiled, and raw linseed oil, cold-pressed and raw walnut oil, and poppy oil from several manufacturers, including Kremer-Pigmente, Fezandie and Sperrle, and Winsor and Newton.

40. Analyzed by the author on 7 and 31 July 1991, using Hewlett Packard HPLC equipped with a diode array detector, National Gallery of Art, Washington, scientific research department, chromatogram files 7_09 and 7_31.

41. Cennini 1966, 46.

42. Analytical results from eight colored paint samples, Susana M. Halpine, report on Agnolo Gaddi, *Madonna Enthroned with Saints and Angels*, 6 August 1990, NGA.

43. Amino acid analysis of brocade paint, bole, and surface layer, Susana M. Halpine, report on Gaddi, *Madonna Enthroned with Saints and Angels*, 23 January 1990, NGA; amino acid analysis of sur-

face layer and "brown" spots, Halpine report on Gaddi, *Madonna Enthroned with Saints and Angels*, 5 March 1991, NGA; amino acid analysis of casein wood glue, gesso ground, and various glues added during previous restoration treatments, Halpine, report on Gaddi, *Madonna Enthroned with Saints and Angels*, 1 October 1990, NGA; fatty acid analysis of "drip," surface layer, dark blue glaze, brownish residue over mordant, and mordant, Suzanne Quillen Lomax, report on Gaddi, *Madonna Enthroned with Saints and Angels*, 22 August 1990, NGA; fatty acid analysis of blue paint and melting point analysis of "drip," Lomax, report on Gaddi, *Madonna Enthroned with Saints and Angels*, 17 October 1990, NGA.

44. The red brocade was not analyzed for fatty acids because of sampling limitations. The determination of egg yolk was based on its appearance and its water insolubility during water extraction.

45. David Erhardt, gum analysis of brocade of Gaddi, *Madonna Enthroned with Saints and Angels*, 16 October 1989, Conservation Analytical Laboratory, Smithsonian Institution, Washington.

46. The sample was taken and prepared by Mary Bustin, former painting conservation fellow, National Gallery of Art, Washington.

47. Cennini 1966, 86–87.

48. Cennini 1966, 87.

49. Cennini 1966, 24.

50. Fatty acid analysis, carried out by the National Gallery, London, scientific department, indicated that the red lake glazing was tempered with egg yolk. Mary Bustin, *Fragments from Two Altarpieces: The Materials and Methods of Lorenzo Monaco*, Courtauld Institute project (London, 1985), 35–36.

51. Robert Feller, "Studies on the Darkening of Vermilion by Light," *Report and Studies in the History of Art*, 1967, National Gallery of Art [1] (Washington, 1967), 107–109.

52. Cennini 1966, 24.

53. A test panel was prepared with gold leaf by Steve Wilcox, frame conservator, National Gallery of Art, Washington; part of it was painted with glue-bound vermilion, another with egg-bound vermilion. Both sides were then coated with egg yolk mixed with a red lake pigment. Lines were scraped away with a small pointed stick to duplicate the sgraffito. The handling of the sgraffito was very different for the two different test areas. On the side with the glue underpaint, there was much more control over the depth of the line, and the resulting sgraffito lines were much smoother. Additionally, by penetrating the top egg layer alone and leaving the glue for later reworking, a "sketch" of the stencil pattern can be laid out and later reworked in detail. In fact, such a "sketch" seems to appear on Gaddi's paint surface at the edges of the red brocade, under the side columns (see fig. 10) and may not have been reworked because it was hidden from view. On the other hand, the side of the test panel with double layers of egg yolk seemed to stretch and tear, creating a more jagged line, and handled more like removing a latex paint. This very subjective test indicated that there may have been additional advantages to the glue underlayer in sgraffito work, where too much pressure could damage the gold leaf layer below it.

54. Cennini 1966, 20.

55. For general recommendation of egg yolk for panel paintings, see Cennini 1966, 91. Cennini 1966 recommended size for blue pigments on panel paintings: 68, dark blues; 36, azurite and Baghdad indigo; and 87, for a blue sgraffito, apply a coat of lead white–egg yolk over the gold, then "temper your ultramarine blue with a little size and a little yolk, perhaps two drops." He also recommended size for some yellow pigments: 29, orpiment; and 30, saffron.

Cennini 1966 recommended size for certain green pigments: 33, verdigris on panel; 32, orpiment and indigo for green; and 33, ultramarine and orpiment for green. He also recommended egg yolk for other green pigments: 31, malachite; 32, azurite and giallorino (yellow) for green; and 34, terre-verte and lead white on panel.

Cennini 1966 suggested a choice of binders for some colors: 24, vermilion, "according to the situations in which you have to use it"; and 36, azurite.

56. Halpine 1992, 22–38.

57. The attribution of the *Judith and Holofernes* panel painting to Andrea Mantegna has been disputed. Keith Christiansen, curator, Metropolitan Museum of Art, New York, suspects that it is the work of an early sixteenth-century miniaturist, possibly Pietro Guindaleri, who was associated with Mantegna and worked from 1464 to 1506. "Some Observations on Mantegna's Painting Technique," in Suzanne Boorsch, Keith Christiansen, David Ekserdjian, Charles Hope, David Landau, et al., *Andrea Mantegna* [exh. cat., Royal Academy of Arts] (London, 1992), 435. There is no dispute, however, about the egg tempera binder of the painting, and therefore the comparison of the binders of *Judith and Holofernes* and *Christ Child Blessing* is still valid.

58. Susana M. Halpine, report on Andrea Mantegna, *Judith and Holofernes*, 8 January 1990, NGA; Halpine, report on Mantegna, *Christ Child Blessing*, 23 January 1990, NGA; samples were taken with the cooperation of Barbara Berrie and Paula DeCristofaro, formerly of the painting conservation department.

59. Andrea Rothe, painting conservator, J. Paul Getty Museum, Malibu, Calif., personal communication, December 1989.

60. Suzanne Quillen Lomax, report on Mantegna, *Christ Child Blessing*, 27 November 1991, NGA.

61. Barbara Berrie, personal conversation and report on Mantegna, *Christ Child Blessing*, 19 May 1992, NGA. The cross section and photograph are by Berrie.

62. Cennini 1966, 103–104.

63. Cennini 1966, 103.

64. Personal conversation with Berrie regarding her report on Mantegna, *Christ Child Blessing*, 19 May

1992, NGA.

65. Cennini 1966, 104–105.

66. Cennini 1966, 105.

67. Christiansen, in Boorsch et al. 1992, 69, 70, 78.

68. Christiansen, in Boorsch et al. 1992, 69, 70.

69. Carol Christensen, "The Painting Materials and Technique of Paul Gauguin," *Conservation Research*, National Gallery of Art, Studies in the History of Art, 41, Monograph Series II (Washington, 1993), 63–103.

70. Mills and White 1987, 142; Merryl Johnson and Elisabeth Packard, "Methods Used for the Identification of Binding Media in Italian Paintings of the Fifteenth and Sixteenth Centuries," *Studies in Conservation* 16 (1971), 145–164.

71. Skans and Michelsen 1986, 63–71; Mills 1987, 85; Lomax, report on fatty acid analysis, NGA; 2 April 1992; Lomax, report on analysis of casein, 7 August 1990, NGA.

72. Susana M. Halpine, report on Paul Gauguin, *Fatata te Miti* (By the sea), oil on canvas, 26 July 1990, NGA. Samples were taken by Carol Christensen, painting conservator, National Gallery of Art, Washington.

73. Susana M. Halpine, report on Gauguin, *Te Pape Nave Nave*, 26 July 1990, NGA. The sample was taken by Carol Christensen. *Te Pape Nave Nave*, National Gallery of Art, Washington, Collection of Mr. and Mrs. Paul Mellon.

74. Susana M. Halpine, report on Gauguin, *Madame Alexandre Kohler*, 4 April 1989, NGA. The sample was taken by Carol Christensen. *Madame Alexandre Kohler*, National Gallery of Art, Washington, Chester Dale Collection.

75. Judith Brodie, "Sir Edward Coley Burne-Jones," in *Art for the Nation* [exh. cat., National Gallery of Art] (Washington, 1991), 160.

76. Susana M. Halpine, report on Edward Coley Burne-Jones, *Saint Barbara*, mixed media on paper, 16 March 1990, NGA. The samples were taken by Shelley Fletcher, head of the paper conservation department, National Gallery of Art, Washington.

77. As mentioned in the discussion of the 1h extraction procedure, the supernatant can be easily distinguished from the precipitate by the higher amount of norleucine, which is the internal standard.

78. David Erhardt, David von Endt, and Walter Hopwood, "The Comparison of Accelerated Aging Conditions through the Analysis of Extracts of Artificially Aged Paper," *American Institute for Conservation Preprints*, 15th Annual Meeting (Washington, 1987), 43.

79. Skeist 1977, 149.

80. Suzanne Quillen Lomax, reports on Burne-Jones, *Saint Barbara*, 20 December 1989, 14 September 1990, and 24 February 1992, NGA.

81. Lomax, report on Burne-Jones, *Saint Barbara*, 24 February 1992, NGA.

82. Lomax, reports on Burne-Jones, *Saint Barbara*, 20 December 1989, 14 September 1990, and 24 February 1992, NGA.

83. Susana M. Halpine, report on Henri de Toulouse-Lautrec, *Maxime Dethomas* and *Lady with a Dog*, 8 January 1990, NGA. The samples were taken by Teresa Longyear, painting conservator, National Gallery of Art, Washington.

84. Teresa Longyear, personal communication, 1989, pointed out the burlap imprinting and the artist's use of these areas for unusual textural effects.

85. Michael Palmer, conservation scientist, fiber analysis in report on Toulouse-Lautrec, *Maxime Dethomas* and *A Lady with a Dog*, 7 November 1990, NGA.

86. Suzanne Quillen Lomax, report on Toulouse-Lautrec, *Maxime Dethomas* and *Lady with a Dog*, 1 August 1990, NGA.

87. The cardboard support was extracted in four steps in order to clarify the sizing mixture: (1) water was added for 15 minutes, then removed for analysis; (2) water was added for 1 hour in an ultrasonic bath, then removed for analysis; (3) a solution of 50% trifluoroacetic acid, 2% beta mercaptoethanol in water, was added while sonicating for 1 hour in an ultrasonic bath, then removed and analyzed; and (4) the remaining precipitate was also analyzed. This more elaborate extraction procedure confirmed the presence of two proteinaceous materials—a collagenous glue, which dissolved readily in the water extractions, and a less soluble unknown proteinaceous material, which both dissolved in the acid solution and partially remained with the paper fibers in the precipitate. The composition of the latter did not match that of casein but was much closer to the composition of egg and, therefore, may be an albuminous protein. These extractions were run by the author on 3 February 1992, data file numbers 1_28B51A.D–1_28E54A.D, NGA.

88. Composition of serum albumin derived from amino acid sequence was provided by the National Biomedical Research Foundation, Georgetown University Medical Center, Washington.

89. William D. Detlefson, adhesives chemist, Adhesives and Chemicals Division, Borden Chemicals, personal conversation, 4 June 1992.

90. Paul J. Angelillo, conservation scientist, report on Toulouse-Lautrec, *Lady with a Dog*, 21 October 1986, NGA; Lisha Glinsman, scientist, report on Toulouse-Lautrec, *Maxime Dethomas*, 8 August 1988, NGA.

91. Farbenindustrie Aktiengesellschaft, "Procédé de collage," French pat. 796,389, 22 January 1936.

92. Leonard B. Lane and John J. Frendreis, "How Blood Adhesives Are Used," *Adhesives Age* 4, no. 5 (May 1961), 20–24; Skeist 1977, 181.

93. Irving Skeist, *Handbook of Adhesives* (New York, 1962), 158. See also Alfred C. Lindauer, "Blood-Albumin Glue," U.S. pat. 1,459,541, 19 June 1923: "I am aware that mixtures of blood albumin, water, lime, ammonia, etc., in different proportions have

been used in the industries for years"; William Allen Drushel, assignor, Haskelite Manufacturing Corp., "Process of Making Waterproof Blood Glue," U.S. pat. 1,476,805, 11 December 1923: "In preparing this [blood] glue it was customary, prior to my invention . . ."; Skeist 1977, 182–187.

Beef blood from slaughterhouses was the primary source for blood glues, while hog and sheep blood could be added as supplements but were not recommended for use by themselves. See Skeist 1962, 159; Carl E. Drugge, "Glue Mix," U.S. pat. 2,963,454, 6 December 1960.

In the late 1880s it is likely that fresh blood rather than soluble dry blood would have been used in manufacturing, as the solubility of albumin rapidly decreases with age; the blood-drying technologies seem to have been developed somewhat later, and, even if available, processing always added to the cost. See Harold Wade, "Improved Process of Desiccation and Apparatus Thereof," Great Britain (G.B.) pat. 112,965, 29 January 1918; Adrianus J. L. Terwen and Cornelis J. C. van Hoogenhuyze, "Improved Manufacture of Decolorised, Odourless, and Tasteless Albumins from Blood," G.B. pat. 123,971, 15 May 1919; Frederick W. V. Fitzgerald, "Improvements in Processes for Preserving the Fluidity of Blood," G.B. pat. 29,896, 9 November 1921; Gustave Z. Nafilyan, "Improvements in Drying or Removing the Liquid from Solutions or Suspensions of Substances in Liquids," G.B. pat. 221,598, 18 September 1924.

The blood supply had to be near the manufacturers. Robert Herman Bogue, *The Chemistry and Technology of Gelatin and Glue* (New York, 1922), 344: "If fresh blood is to be used the supply must be readily accessible to the consuming plant"; John M. Gossett, Max H. Estep Jr., and Mahlon J. Perrine, "Production of Glue from Soluble Blood," U.S. pat. 2,874,134, 17 February 1959.

94. William T. Brannt and William H. Wahl, *The Techno-Chemical Receipt Book* (Philadelphia, 1887), 324–325; Lane and Frendreis 1961, 20–22.

95. Nikolaus Prevner, "The Early History of Plywood," *Architectural Review* 86, no. 514 (September 1939), 129–130: Russian plywood manufacture was "successful, above all—and this is the next important step in the history of plywood—when he had introduced a new waterproof casein glue (1892)"; Alexander W. Katlan, *American Artists' Materials: A Guide to Stretchers, Panels, Millboards, and Stencil Marks* (Madison, Conn., 1992), 2:42: "By 1830, the German manufacturer, Michael Thonet, was producing plywood furniture, which he showed in the Crystal Palace Exhibition of 1851"; Terry Sellers Jr., *Plywood and Adhesive Technology* (New York, 1985), 7–8: "In 1830 the piano industry began using plywood, perhaps the first industry in the United States to do so. . . . In 1905 at the Lewis and Clark Exposition of the World's Fair held in Portland, Oregon, the first construction plywood composed of three-ply Douglas fir was made and displayed."

96. Drushel 1923, 1; Lindauer 1923, 1; Don Brouse, "Behavior of Casein and Blood Glue Joints on Exposure," *Wood Products* (Chicago) 12 (October 1934), 8–10; Charles B. Norris, *Technique of Plywood* (Seattle, 1942), 229–230; A. M. Kragh and J. Wootton, "Animal Glue and Related Protein Adhesive," in *Adhesion and Adhesives*, ed. R. Houwink and G. Salomon, 2d ed. (Amsterdam, 1965), 1:164; Bogue 1922, 347; Skeist 1977, 181.

97. A. Luther, German pat. 198,182, 13 August 1905; Arie Tinkelenberg, Henricus W.L.M. Vaessen, Kwai W. Suen, and Anton J. Van Doorn, "Manufacture of Chipboard Having High Strength and Reduced Formaldehyde Emission, Using a Minor Amount of Protein in Combination with Low Formaldehyde-Urea Resins," U.S. pat. 4,282,119, 4 August 1981; Skeist 1962, 168–169.

98. Fritz Lux, "Process of Producing a Cold Glue from Blood," U.S. pat. 1,468,313, 18 September 1923; Henry V. Dunham, "Glue and Method for Making Plywood," U.S. pat. 1,892,486, 27 December 1932; Norris 1942, 230; Lindauer 1923, 1; Stanislao A. De Santis, organic chemist and retired manufacturer of blood glue for U.S. plywood industry, personal conversation, Washington, 10 September 1990. De Santis said he had been approached by cardboard manufacturers requesting blood glue, but the production was unable to meet the increased demand.

99. "Blood-Glue Solves the Problem," *Western Flyer* 3 (October 1927), 84; Skeist 1977, 181; Henry L. Haskell, assignor, Haskelite Manufacturing Corp., "Laminated Material and the Process of Preparing Same," U.S. pat. 1,516,566, 25 November 1924.

European use of blood glue continued to increase, and the demand in the American plywood industry peaked in 1965. Skeist 1962, 168: "Blood glues contribute to the over-all scheme of wood utilization in such a variety of unique and important ways that their future seems closely related to that of the forest products industry itself."

William D. Detlefson, "Blood and Casein Adhesives for Bonding Wood," *Adhesives from Renewable Resources*, American Chemical Society 1987 Symposium in New Orleans, ser. 385 (Washington, 1989), 446. After 1931 phenol-formaldehyde and urea-formaldehyde resin film glues first diluted and eventually replaced blood glues for most applications in the United States. Detlefson, personal conversation, 1992; Skeist 1977, 181; Adolf Menger, "Adhesive," U.S. pat. 2,203,501, 4 June 1940: blood albumin and formaldehyde-urea adhesive; Carl S. Leonardson and Donald J. White, "Animal Protein Resin Product and Preparation Thereof," U.S. pat. 2,332,802, 26 October 1943; Charles N. Cone, "Resin-Blood Glue and Process for Making the Same," U.S. pat. 2,895,928, 21 July 1959: blood glue and phenol-formaldehyde adhesive; Frederick James Shelton and Charles H. Chervenka, "Plywood Adhesive Comprising Urea Formaldehyde Resin Protein and Dispersing Agent and Process for Preparing Same," U.S. pat. 2,872,421, 3 February 1959; Tinkelenberg et al. 1981.

100. James P. Casey, *Pulp and Paper* (New York, 1952), 2:1232.

101. John Gray, "Improvements in and Relating to Cements Impervious to the Action of Water," G.B. pat. 202,404, 20 August 1923.

102. Some of the blood glue patents that specifically mention paper, cardboard, pressboard, or paperboard as an application are Gray 1923; John Gray, "An Improved Cementing Composition for Use in Preparing Laminated Materials," G.B. pat. 366,889, 11 February 1932; Albert Henry Bowen, "Adhesive and Process of Making Same," U.S. pat. 2,014,167, 10 September 1935; Lawrence Bradshaw and Charles Logan Stachel, "Process of Making Adhesive Material," U.S. pat. 2,391,387, 25 December 1945; Norman J. Sheeran, "Laminating Adhesive and Method of Producing the Same," U.S. pat. 2,870,034, 20 January 1959; Carl E. Drugge and John M. Hine, "Glue Mix," U.S. pat. 2,963,454, 6 December 1960; Ambros A.G., "Procédé de préparation de produits de condensation solubles dans l'eau à partir de substances albumineuses et d'acide sulfonique de lignine," French pat. 1,438,765, 4 April 1966; Sulo A. Karjala and Frank K. Dering, "Process of Reacting Blood Proteins with Glutaraldehyde and the Product Thereof," U.S. pat. 3,294,564, 27 December 1966; Sulo A. Karjala and Frank K. Dering, "Process of Treating Proteins and the Product Thereof," U.S. pat. 3,301,692, 31 January 1967; Roy E. Wright and Joseph Vastag, "Adhesive for Water-Resistant Corrugated Paperboard Bonds," U.S. pat. 3,704,148, 28 November 1972; C. Karl Bunnig, "Verfahren zum Herstellen eines Klebstoffes, der als Bindemittel bei der Herstellung von Holzwerkstoffplatten bzw. beim Verkleben von Holz, Platten, Papier oder anderen cellulosehaltigen Naturprodukten verwendbar ist," German pat. 2,715,501, 12 October 1978; Bunnig, "Method for Manufacturing Glue for a Binding Agent in Manufacture of Chipboard or Other Cellulose Products," U.S. pat. 4,180,412, 25 December 1979.

103. Ambros A.G. 1966.

104. Bunnig 1979.

105. The author used a phenolphthalein presumptive test and a Takayama confirmatory test seeking evidence of hemoglobin, following procedures in the *FBI Laboratory Serology Unit Protocol Manual*, Federal Bureau of Investigation, U.S. Department of Justice (Washington, 1989), 2–3, 2–4, 2–10, 2–11; Robert Spalding, head of Serology Unit, Federal Bureau of Investigation, Washington, personal communication, 13 September 1991; De Santis 1990; Charles N. Cone, "Glue and Process of Manufacture Thereof," U.S. pat. 2,400,541, 21 May 1946; Lux 1923, 1; Lindauer 1923, 1; Skeist 1977, 183. De Santis mentioned heating and the addition of sodium citrate and sodium hydroxide for drying and maintaining a high pH in blood glue production, a treatment that Spalding speculated might have contributed to the deterioration of the hemoglobin used for blood identification.

106. Halpine 1992, 22–38. Cosimo Tura's use of egg tempera and distemper on the same panel painting is discussed.

107. Susana M. Halpine, report on organic paint analysis of Toulouse-Lautrec, *Maxime Dethomas* and *A Corner of the Moulin de la Galette*, 8 January 1991, NGA; Halpine, report on fatty acid analyses done on HPLC, 9–12 December 1991, NGA.

APPENDIX

Glossary of Technical Terms

Accelerated (or artificial) aging: process whereby samples are aged by placing them in a fadeometer or weatherometer (Atlas) instrument and exposing them to a xenon arc lamp simulating daylight, at 55°C and with 50 percent relative humidity.

Amino acid analysis: protein characterization technique that breaks the bonds between the constituent amino acids in a protein sample, measures the amounts of free amino acids, and gives the percent composition of the amino acids that make up that protein.

Gas chromatograph (GC): an analytical instrument similar to an HPLC, using a gas to carry the analizate through a column for separation. Several detectors can be installed after the column, including a flame ionization detector or a mass spectrometer (GC–MS), both of which are installed in the National Gallery of Art, scientific research department. The GC is used to analyze a wide range of materials and has been used for a number of years in conservation science to identify oils, waxes, resins, and other organic materials.

High-performance liquid chromatography (HPLC), also called high-pressure liquid chromatography: an analytical technique in which a material, once solubilized, is separated from other materials by being carried along by a solvent stream through a particle bed within a packed column and selectively interacting with a matrix contained in the column. The ability to manipulate the solvent system and column packing, in addition to factors such as flow rate, temperature, and choice of detectors, gives the instrument its flexibility in analyzing materials ranging from hydrophobic materials such as oils and resins to hydrophilic materials such as proteins and carbohydrates. HPLC instruments are beginning to be used in conservation for amino acid analysis and dye analysis.

PITC: Phenylisothiocyanate, a reagent used during the derivatization step to label the hydrolyzed amino acids. PITC functions as a chromophore, tagging free amino acids and absorbing at a higher wavelength (254 nanometers) than unlabeled amino acids.

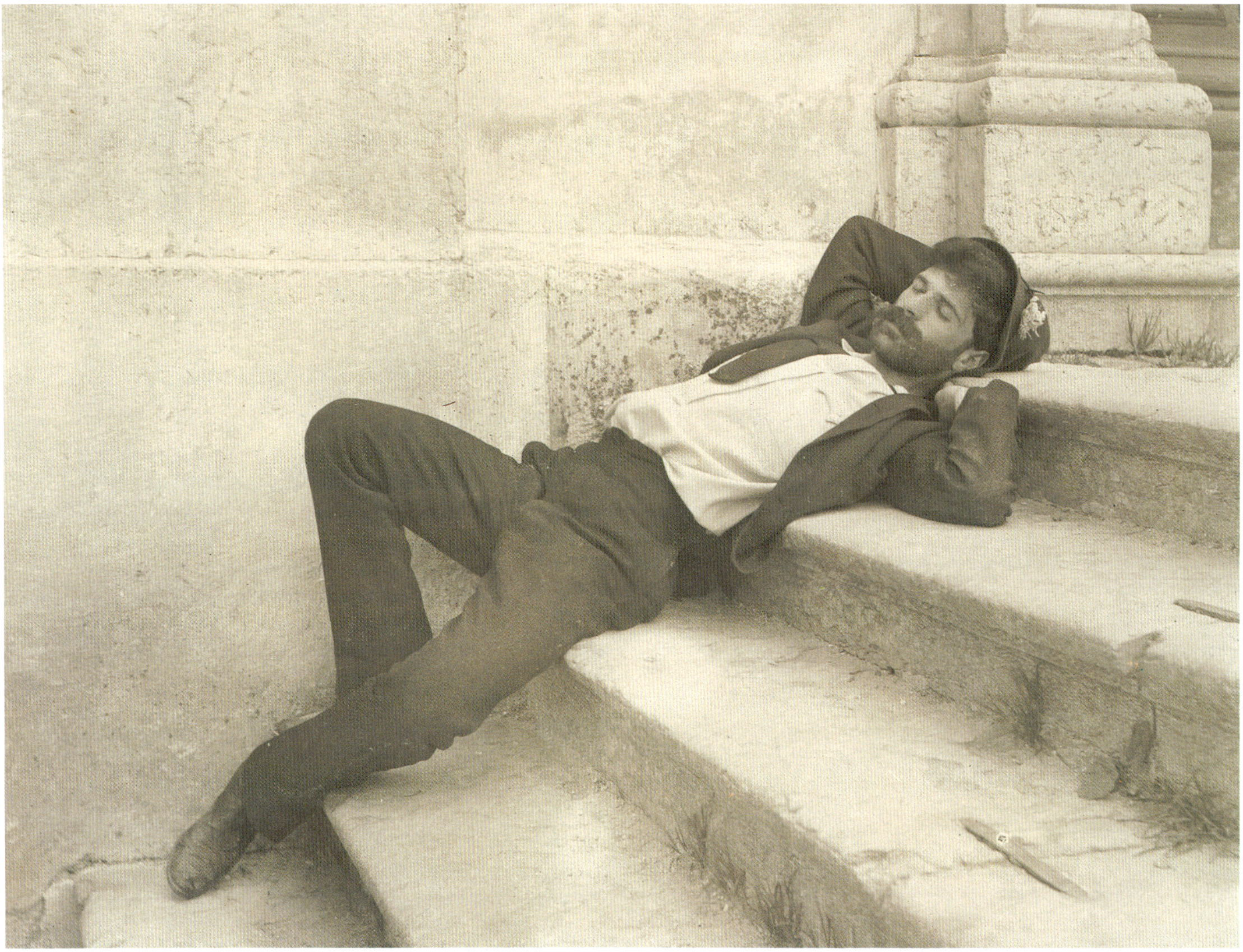

CONSTANCE McCABE and LISHA DEMING GLINSMAN

Understanding Alfred Stieglitz' Platinum and Palladium Prints: Examination by X-ray Fluorescence Spectrometry

Alfred Stieglitz (1864–1946) was one of the leading experimenters and innovators in photography, known for his aesthetic vision and highly respected as a photographic chemist. At the turn of the century, he published extensively on the platinum printing process and its many variants. He used this process almost exclusively until the increase in the price of platinum during World War I made platinum papers nearly unobtainable. The unavailability of platinum gave rise to an alternative process, which substituted salts of palladium for salts of platinum, to produce the light-sensitive photographic paper.

Stieglitz' technical publications describe many of his early approaches to the platinum process, including generic recipes for sensitizers, developers, and toners and other variations in his darkroom techniques.[1] Unfortunately, he published nothing regarding his use of the palladium process, and mentions of it in his personal writings are ambiguous.[2] Despite his training as a scientist, it appears that Stieglitz did not systematically record his photographic experiments or the specific methods by which his individual prints were made.

Stieglitz' published literature is a valuable aid for understanding the generic approaches he used to produce his platinum photographs. However, that no other records exist regarding his extant platinum or palladium prints is unfortunate from the perspective of a photograph conservator who must deal with a photograph as a physical object made of a complex component structure. Unlike some other processes, platinum prints, palladium prints, and variants of these processes are frequently indistinguishable by their appearance alone; the chemical nature of Stieglitz' prints simply cannot be accurately or completely described by visual examination (fig. 1).

1. Alfred Stieglitz, *Self-Portrait, Cortina*, 1890, platinum print, probably processed with a mercury developer
National Gallery of Art, Washington, Alfred Stieglitz Collection 1949.3.63

The National Gallery of Art, Washington, is fortunate to house the largest existing collection of photographs by Alfred Stieglitz, selected from his estate by Georgia O'Keeffe as the "key set" of his finest finished work. Responsible approaches must be taken to preserve these artistically and historically important photographs by maximizing their longevity and continued availability for viewing without significantly altering their physical characteristics. To accomplish this goal a conservator must be familiar with the artist's aesthetic intent. Equally important, however, is an understanding of the physical and chemical nature of the photograph to be treated or displayed. A reasonable understanding of the chemical and physical nature of any work of art or historical artifact is essential for making decisions regarding its preservation and exhibition needs and for approaching conservation treatment. Disastrous fading or darkening could result, for example, from the exhibition of a light-sensitive photograph that is misidentified as one that is stable to light.[3]

Secondary emission energy-dispersive x-ray fluorescence spectrometry (XRF) has been used with some success in the past to gain a better understanding of the chemical nature of the final image in photographs.[4] The following are results of new research further developing an analytical protocol designed specifically for the use of XRF in studying photographic image materials. This new protocol has been successfully used to determine what metallic elements make up the final image in some of the photographs in the collection of the National Gallery of Art. The resulting data have served to extrapolate methods of production that Stieglitz might have used to fabricate his platinum and palladium photographs.

The Platinum and Palladium Printing Processes

The platinum printing process was favored by many turn-of-the-century photographers, including Stieglitz. William Crawford's excellent description of the process in *The Keepers of Light* is indispensable in assisting modern practitioners to recreate early photographic processes.[5] The first step in producing a photographic paper is to coat it with a sensitizer. Platinum printing paper is sensitized with an aqueous solution of one of the platinous compounds, generally sodium tetrachloroplatinite tetrahydrate, $Na_2PtCl_4.4H_2O$, or potassium tetrachloroplatinite, K_2PtCl_4 (sometimes referred to in the literature as sodium or potassium chloroplatinite, or as sodium or potassium tetrachloroplatinate), and ferric oxalate, $Fe_2(C_2O_4)_3.5H_2O$, which is allowed to dry. This light-sensitive paper is then placed in contact with a negative and exposed to light. The action of light converts the ferric salts to ferrous salts that gradually become visible as a weak printed-out image. The print is then placed in a potassium oxalate developer that serves to dissolve the new ferrous salts, $FeC_2O_4.2H_2O$, thereby reducing the potassium tetrachloroplatinite to metallic platinum. The action of development completes the image-forming process. Any unexposed and soluble ferric salts in the print are then eliminated by converting them to ferric chloride by clearing (or fixing) the print in a dilute hydrochloric acid solution. These soluble iron salts are removed in the final water wash. In other words, this process uses the light-sensitive properties of an iron(III) compound, which, on exposure to light, is reduced to an iron(II) compound, which in turn has the ability to reduce a compound of platinum to its metallic state. When this procedure is properly carried out, the final image of a platinotype consists of chemically stable, finely divided metallic platinum.

The primary difference between palladium prints and platinum prints is that palladium salts replace the salts of platinum in the sensitizer. The palladium compounds generally used to sensitize the paper are in the form of potassium tetrachloropalladite, K_2PdCl_4, or sodium tetrachloropalladite trihydrate, $Na_2PdCl_4.3H_2O$. The chemical processing of palladium prints is similar to that of the platinum process, but the resulting final image consists of metallic palladium.[6]

Because of the similarity of the chemical processes used to produce platinum and palladium prints, the variants of both processes are also virtually interchangeable. These variants can include numerous chemical formulations of sensitizers, developers, and toners; variations of the length of a print's exposure; or variations of the developing solution temperature. The variants of both processes were used to manipulate the image hue and tonal properties. Stieglitz continually experimented with these chemical variants to achieve very specific aesthetic results. While he mastered both processes, he considered the platinotype to be "that prince of all processes,"[7] and it took years for him to accept the palladium process as a viable substitute for platinum.

Platinum prints tend to have a neutral gray-black image hue, while palladium prints are generally slightly warmer in hue. However, depending on variations in processing, such as temperature of the developer, the image hues of either process can be manipulated to range from cool, slightly purple-black to a very warm brown. Modifying standard processing procedures, such as altering the formulation of the sensitizer or developer or employing chemical toning or intensification, can result in a wide variety of hues. One possible modification of either process is toning the finished print with gold. Gold toning can change a warm image to black or purple-black. Intensifiers, such as a combination of gallic acid and silver nitrate, have been used to change the image color, increase the image contrast, and highlight the detail of a finished print. In addition, any of these modifications can have an impact on the chemical stability of photographs and result in changes in the appearance of prints over time. For example, toning with compounds such as uranium can change the color of the image to red, green, or blue or can even result in split tones in a print. However, uranium toning is known to result in gradual changes in the appearance of a finished print. By 1911 it was known that the use of uranium as a toner "does not give very permanent results."[8]

To Stieglitz, an important feature of the basic, unmodified platinum process was its permanence when compared, for example, to the silver processes, such as salted paper prints, albumen prints, or gelatin prints. In a 1902 article he said, "We hear much of the permanence of the platinum print, and while it is undoubtedly true that a careful and conscientious technical handling of the process will insure a print practically as permanent as its paper support, yet with modern sloppiness this is no longer true."[9] He acknowledged, however, the value of other, less permanent processes: "I willingly grant that 99 per cent. of the negatives produced by the 'camera artists' ought to be printed on a most *non*-permanent process, for the sake of coming generations."[10]

Identification of Photographic Processes

It is frequently possible to distinguish the process by which a photograph was made just by looking at it. In general, an experienced observer can recognize the difference between an albumen print and a gelatin print and can further differentiate between a printed-out versus a developed-out gelatin print.[11] Some photographs, however, are difficult or impossible to identify visually in terms of process. Single-layer photographs in which the final image exists in the upper surface of the paper support, such as platinum and palladium prints, can be particularly difficult to identify.

As discussed, varying the chemical formulations used to produce platinum and palladium prints results in a variety of image appearances; therefore, visual identification is complicated. For example, platinum prints (considered cool or neutral in hue) can be made warm by the addition of a mercury compound to the developer. Conversely, a warm hue is usually associated with palladium prints. Cooler hues, however, can be attained by developing or toning the print in various chemical solutions or at different temperatures, as described above. Because of these variations, platinum prints sometimes masquerade as palladium prints, and vice versa. In some cases, a salted paper print might be confused with a platinum or palladium print. Further complicating the issue, many silver gelatin prints were manufactured to mimic the appearance of platinum papers and other historic process types. In addition, visual examination alone cannot determine what chemical toner might have been used to alter the original appearance of a print.

Photographs that have an ambiguous appearance require more than visual examination to determine the components that make up the final image. The influence of so many factors produces an almost infinite variety of nuances and subtleties in the appearance of different types of photographs. It is this enormous variety of processes and final appearances that has led even experienced observers to turn to chemical analysis, such as examination by XRF, for objective information. This information is especially useful for prints made by an unknown photographer or by an experimenter who kept no records, such as Alfred Stieglitz.

XRF is a well-established analytical technique that provides rapid, simultaneous multielement analyses in many applications. Because of its nondestructive capabilities, it has proven valuable in the study of art objects. To use XRF for the study of photographs of unknown composition, it is important first to determine if the very small traces of elements that make up the photographic image can, in fact, be detected by the instrument.

Fabrication of Photographic Facsimiles for Use as XRF Standards

In using XRF to analyze photographs, it is essential to maximize the sensitivity of the instrument to any relevant element that might occur in the final image material. For this reason, facsimile photographs of known composition were produced and analyzed by XRF, and the detection of the known image materials was optimized. It was determined that the elements that made up the image in the known samples could, indeed, be detected by XRF. As a result, a protocol was developed that used the facsimiles produced by known chemical processes as analytical standards. These standards were compared with several Stieglitz photographs in the National Gallery of Art.

To assist in producing analytical standards with which to compare Stieglitz' prints, photographic working methods of Stieglitz and his contemporaries were studied. Stieglitz generally used a potassium oxalate solution as a developer for his platinotypes, to which he sometimes added glycerine to delay the action of the developer and a mercury compound applied locally to modify the image hue.[12] He was also known to use uranium compounds to tone his platinum prints.[13] Salts of gold and even iridium and osmium were also mentioned in contemporary handbooks for use as toners for platinum prints.[14] Therefore, the presence of any of these elements could be attributed to the developer or to a postprocessing toning treatment.

Early in his career, Stieglitz experimented with sensitizing his own platinum paper. For most of his platinum work, however, he used commercially produced, presensitized photographic papers. He carried out his chemical processing experiments using commercial papers, including those made by Eastman Company (American Platinum Paper),[15] the Platinotype Company, and Willis and Clements.[16] While Stieglitz described the inclusion of "lead chloride or a mercury salt to the sensitizer" in a patent by W. Willis as an improvement of the process,[17] very little is known about what papers were used to produce specific prints. Because commercially manufactured papers were made with a variety of chemical sensitizer formulations, it is possible that along with metallic platinum or palladium, lead or mercury might be present as part of the final image.

Hydrochloric acid was generally used for the clearing or "fixing" baths, followed by a running water wash. If fresh baths of hydrochloric acid were used until the print was thoroughly cleared, followed by a water wash that thoroughly removed residual processing chemicals, the image in the final print should consist of metallic platinum and/or palladium possibly in combination with traces of mercury, uranium, or other elements that were deposited on the primary image due to chemical reactions of the sensitizer, along with the developer and possible toners.

There are many ways to produce platinum and palladium prints and their variants. However, the basic recipes used for sensitizers, developers, and toners are similar in many handbooks and journals of Stieglitz' time. For this reason, facsimiles of processes that may have been used by Stieglitz were produced by the author (McCabe) according to recipes given by Crawford in *The Keepers of Light*.[18] These recipes were selected because they are chemically similar to those given in early descriptions by Stieglitz.[19] Although Stieglitz did not include sodium sulfite in his 1893 recipe for uranium toner,[20] Crawford's recipe is virtually identical to one described in 1902 by James McCorkle in *Camera Notes*,[21] both of which refer to the crystal of sodium sulfite to be used as the size of a "pea." The photographic facsimiles were employed as XRF standards for comparison with the Gallery's Stieglitz Collection photographs. They are:

- platinum print
- palladium print
- combination platinum-palladium print (50:50 ratio)
- palladium print, mercury developed
- combination platinum-palladium print, mercury developed
- palladium print, gold toned
- palladium print, silver intensified
- platinum print, uranium toned

X-ray Fluorescence Spectrometry and Photographs

The interpretation of the XRF spectrum of a photograph can be better understood with some basic information on how XRF works. XRF uses x-rays for sample excitation. X-ray photons that are directed at a sample, such as a photograph, interact with atoms in the sample. As an x-ray travels through a sample, it may be absorbed, giving up its energy to a sample electron that escapes from its

vertical scale=2,000

2. XRF spectrum of a bronze standard. Note the clear peaks indicating the presence of copper, lead, and tin and the low amount of background "noise." Also note the very small inelastic scatter peaks related to the rhodium and the barium chloride ($BaCl_2$) secondary target

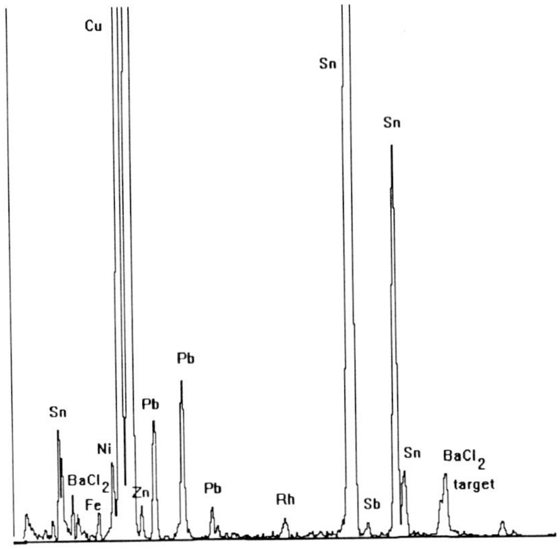

fig. 2 range: 40keV

orbital, leaving a vacancy. The atom is then in an unstable (excited) state. To regain stability, an electron from an outer shell drops into the inner shell vacancy, emitting x-rays as it does so. The atom is said to be in the stable (ground) state. Each x-ray emitted has a specific energy described by the energy difference between the excited and ground states of an atom or ion. These are referred to as transitions. Each element has many possible series of transitions occurring simultaneously. These result in the generation of the K, L, M (among others) transition lines that are observed as peaks on the x-ray spectra. The energies of the electron transitions are characteristic of the elements present in a sample and are detected and recorded as a series of peaks in a spectrum by the XRF instrument.[22]

XRF analyses of photographs are difficult to interpret, due primarily to the low signal-to-noise ratio, which is due to the extremely low physical density of the paper support of the photograph as well as the small amount of metal actually present in a print.[23] When the spectrum for a photograph is compared, for example, to a bronze sculpture, which has a high density, interpretation of the distinct peaks found in the spectral data of the high-density bronze is relatively straightforward (fig. 2). However, the low density of photographs results in a large amount of inelastic scatter in the spectrum when compared to the small amount of metal actually being analyzed. This scatter is seen as a high level of background "noise" that disguises the actual spectral peaks. Because these peaks cannot be positively distinguished from the noise, an XRF protocol was developed to minimize this noise and enhance the spectral lines of specific elements relevant to the photographic materials in question.

The detection of either the two major K transition lines (α and β) or the three major L transition lines (α, β, and γ) was required to confirm the presence of the specific final image constituents. The metals palladium (Pd) and silver (Ag), and impurities such as calcium (Ca), iron (Fe), copper (Cu), and zinc (Zn), are determined by the presence of their series of K transition lines. Platinum (Pt) and traces of elements that might have been used in the toner or developer formulation are identified by the presence of their series of L transition lines. These elements could include mercury (Hg), gold (Au), or uranium (U) (fig. 3a–b).

vertical scale=1,000

3a. Detail, XRF spectrum of fig. 1, seen with a molybdenum (Mo) secondary target. Note the small but well-defined peaks for platinum (Pt) and mercury (Hg)

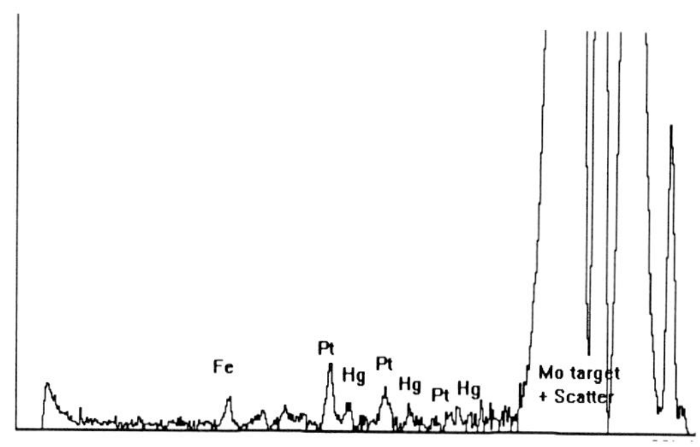

fig. 3a range: 20keV

vertical scale=2,000

3b. Detail, XRF spectrum of fig. 1, with a barium chloride secondary ($BaCl_2$) target. Note the indistinct peaks for platinum and the absence of palladium peaks

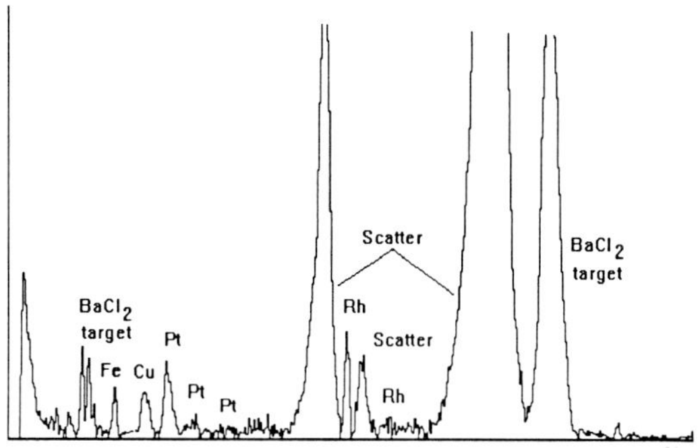

fig. 3b range: 40keV

Developing an XRF Protocol for Analyzing the Elemental Content of Photographic Images

To interpret the XRF spectrum of a photograph and determine the image material, the elements present in the paper support as well as those in the mounting board must be eliminated from the spectrum so that analysis can be based on the contribution from the image components alone. A protocol was developed that acquired a spectrum from a dark region (maximum image density) of the photograph and one from a light region (minimum image density). When possible, a spectrum from the mounting board was also obtained. Once the elemental contributions of the photograph's light region and its mount are subtracted from the spectrum of its dark region, the resulting spectrum will characterize those elements used in the final image material of the printing process. The spectra obtained by analyzing the National Gallery of Art Stieglitz photographs were compared to the spectra of the facsimile photographs (standards) produced according to methodology likely to have been employed by Stieglitz. Note that in some of Stieglitz' prints there is no image area that has a very low-density image (that is, very light). In these cases, the data from the lightest area were acquired, but generally these spectra were not subtracted from the spectra of the maximum density area.

The XRF analyses were performed using a Kevex 0750A spectrometer, which is a free-standing instrument with the x-ray tube and detector mounted on a steel column that allows for vertical and horizontal movement of the spectrometer (fig. 4). The photograph is placed on an easel in front of the XRF where a laser and incandescent lamp are used to focus on the desired sample area. An acrylic block was placed behind the photograph to prevent the x-rays from passing through the photograph to the back of the easel. Since the Kevex 0750A is an air path instrument, it was not equipped with a light-element detector. Therefore, elements lighter than sulfur were not detected with this setup and are not included in these analyses.[24]

The XRF was equipped with six millimeter collimators and either a barium chloride ($BaCl_2$) or molybdenum (Mo) secondary target.[25] Secondary targets provide nearly monochromatic x-rays to excite the sample and enhance the spectral lines of the elements analyzed. When the element's spectral lines are close to (and above) the absorption edge of the secondary target, the enhancement of those lines improves significantly. After considerable experimentation with a variety of XRF conditions and secondary targets, two sets of conditions were found to be the most useful for detecting the metallic elements relevant to platinum and palladium photographs. A barium chloride target was initially used to differentiate the primary metallic elements of the final image material, that is, platinum or palladium. A molybdenum target, which improves the detection of platinum and other relevant metallic ele-

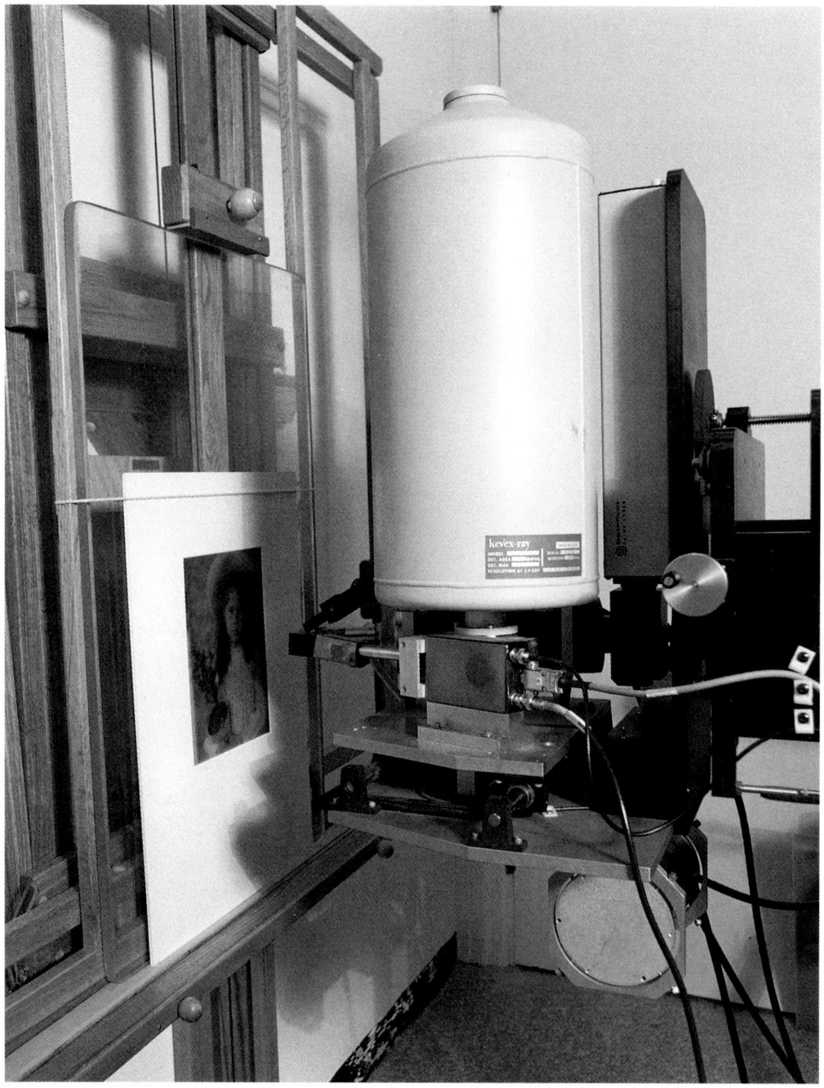

4. The XRF instrument with Stieglitz' *Kitty* (see fig. 15) on an easel prepared for analysis

ments, was then used to detect the presence of any elements that might be attributable to the use of special developers or toners. The palladium K transition lines are higher in energy than the molybdenum lines of the secondary target and will not be excited and, therefore, cannot be seen with the molybdenum target. However, the L transition lines of mercury, gold, and uranium become enhanced and are readily detected when using the molybdenum target.

Results of XRF Analysis of Photographic Facsimiles

The results of XRF analyses clearly show that platinum and palladium photographs can be easily distinguished from one another. Figure 5 is an XRF spectrum of a platinum photographic facsimile using a molybdenum secondary target. Here the platinum peaks are readily evident. Figure 6 is an XRF spectrum of a palladium photographic facsimile using a barium chloride secondary target for the analysis. This spectrum reveals not only the higher-energy K lines for palladium, but also the lower-energy L lines of palladium.

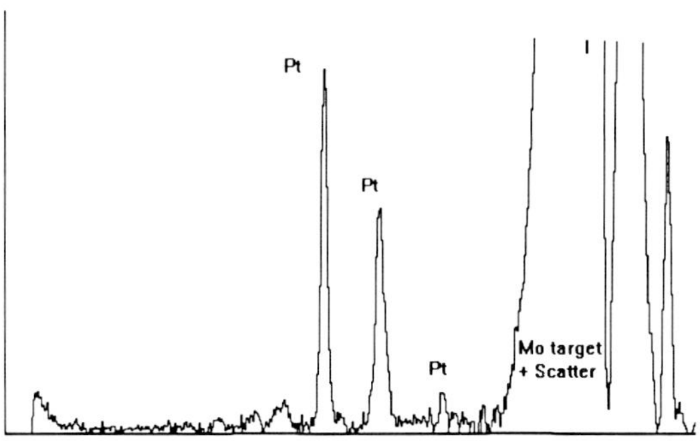

fig. 5 range: 20keV

vertical scale=1,000

5. Detail, XRF spectrum of a dark area of a standard platinum photographic facsimile, with the background subtracted, using a molybdenum secondary target. Note the presence of the three major L transition lines (α, β, and γ) for platinum

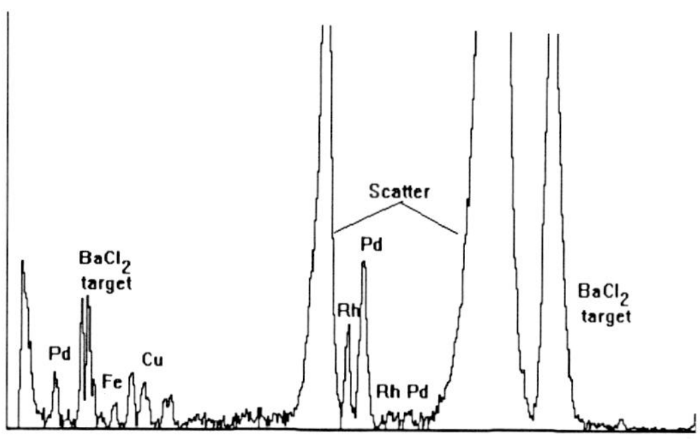

fig. 6 range: 40keV

vertical scale=2,000

6. Detail, XRF spectrum of a dark area of a standard palladium photographic facsimile, with the background subtracted, using a barium chloride secondary target. Note the presence of the two major K (α and β) transition lines for palladium (Pd), as well as the lower-energy L transition lines

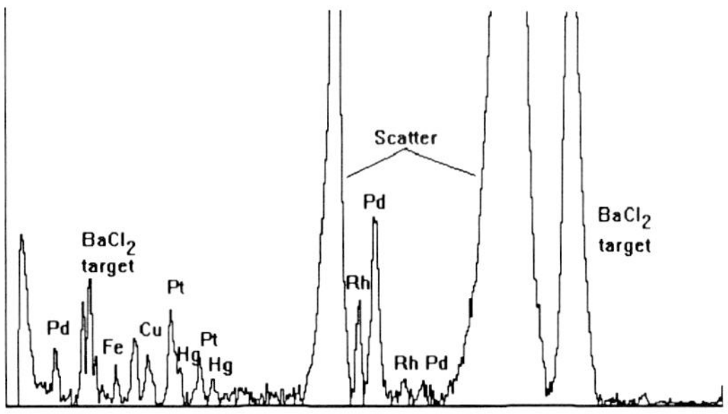

fig. 7a range: 40keV

vertical scale=2,000

7a. Detail, XRF spectrum of a dark area of a standard mercury-developed combination platinum and palladium photographic facsimile, with the background subtracted, using a barium chloride secondary target. While the peaks for palladium (Pd) are well defined, the peaks for mercury (Hg) and platinum (Pt) are not distinct. A molybdenum (Mo) target must be used to enhance the detection of these elements

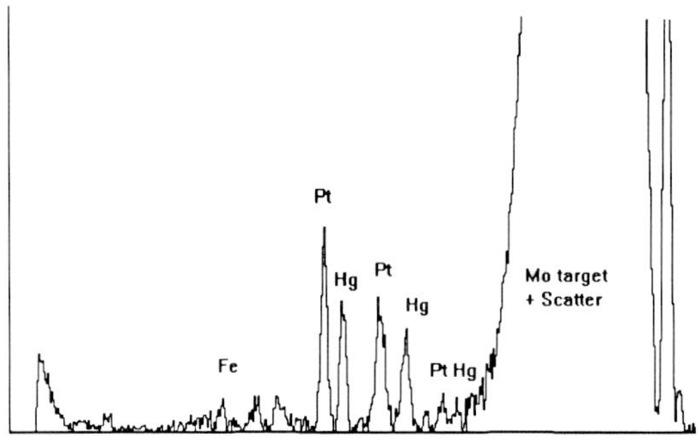

fig. 7b range: 20keV

vertical scale=1,000

7b. Detail, XRF spectrum of same, with the background subtracted, using a molybdenum (Mo) secondary target. The platinum (Pt) and mercury (Hg) peaks are clearly defined

The usefulness of varying the secondary target to optimize the elemental information can be demonstrated with the XRF analysis of a mercury-developed combination platinum and palladium print. In the XRF spectrum with a barium chloride secondary target (fig. 7a), platinum and palladium peaks are easily observed; however, the traces of mercury are barely visible. By using a molybdenum secondary target (fig. 7b), the mercury lines become enhanced and fully resolved from the platinum lines. The palladium K lines are higher in energy than the molybdenum lines of the secondary target and cannot be excited and, therefore, are not detected on this spectrum.

Enhancement of the spectral lines by changing the secondary target is again illustrated by analyzing the standard gold-toned palladium photographic facsimile. The palladium K and L lines are discernible when the barium chloride target is used. However, only the L α line for gold is detected above background. Repeating the analysis with the molybdenum secondary target increases the excitation of the small amount of gold present, causing all three L lines to be clearly visi-

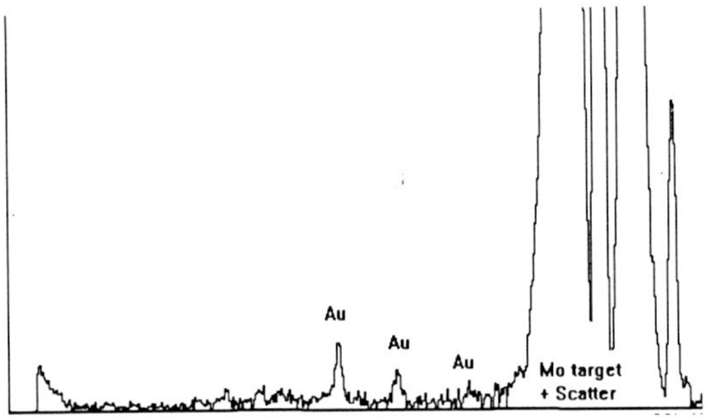

fig. 8 range: 20keV

vertical scale=1,000

8. Detail, XRF spectrum of a dark area of a standard gold-toned palladium photographic facsimile, with the background subtracted, using a molybdenum secondary target. The presence of gold (Au) is detected

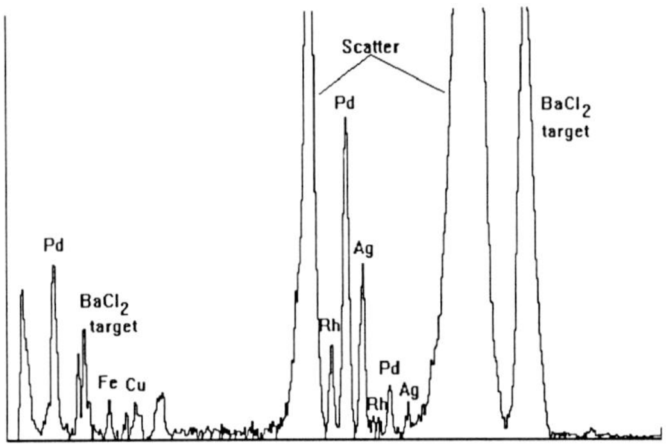

fig. 9 range: 40keV

vertical scale=2,000

9. Detail, XRF spectrum of a dark area of a standard silver-intensified palladium photographic facsimile, with the background subtracted, using a barium chloride secondary target. The presence of silver (Ag) is detected

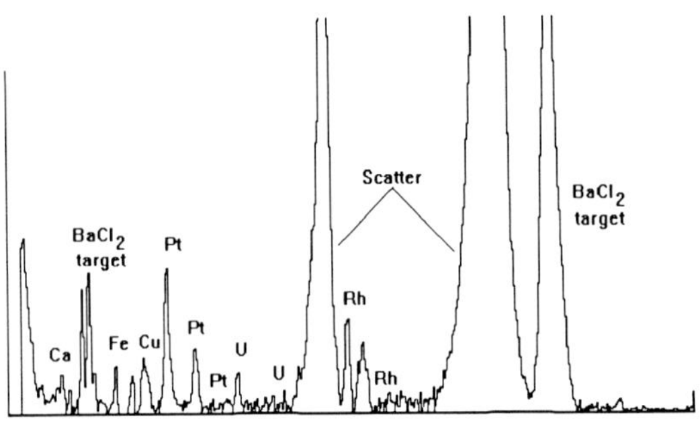

fig. 10 range: 40keV

vertical scale=2,000

10. Detail, XRF spectrum of a dark area of a standard uranium-toned platinum photographic facsimile, with the background subtracted, using a barium chloride secondary target. The presence of uranium (U) is detected

ble (fig. 8). XRF analysis can also be used to detect silver in a silver-intensified palladium print (fig. 9). In the event that uranium toning was used, uranium can be detected (fig. 10). It was determined that to detect all the elements that might make up the final image, and thus to characterize fully each unknown collection photograph, both excitation conditions (using the barium chloride and the molybdenum secondary targets) must be used.

Results of XRF Analysis of Stieglitz Photographs

Several photographs in the National Gallery of Art's collection of Alfred Stieglitz photographs were selected for examination by XRF. Some of these were chosen because of their unusual appearance and others because of skepticism expressed regarding earlier records of process identification from unknown sources. The accurate identification of the processes was a critical element in the intellectual content of the fall 1992 exhibition at the National Gallery, *Stieglitz in the Darkroom*. Therefore, selected photographs were analyzed to establish or verify the identification of the processes by which they were made.

Two portraits of Katharine N. Rhoades, printed from the same negative displayed in *Stieglitz in the Darkroom* (figs. 11–12), had been previously identified by unknown sources as platinum prints. Figure 11 has the archetypal appearance of a platinum print, with a very neutral image hue accompanied by off-white highlights. Figure 12 has a very different appearance; both its image hue and highlights are quite warm. After completing analysis of these photographs using the XRF protocol described above, platinum was found to be present in both the neutral-hued and the warm-hued portraits. However, a trace of palladium was also detected in both portraits (figs. 13–14); the two spectra are nearly identical. It is possible that the presence of palladium contributes significantly to the difference in appearance. However, variations in the temperature of the developer may have caused the differences in image hue. Despite the newly revealed information regarding the final image materials of these prints, the factors that play the primary role in their appearance remain unclear.

OPPOSITE PAGE, LEFT
11. Alfred Stieglitz, *Katharine N. Rhoades at '291'*, 1915/1916, platinum print with a trace of palladium. This photograph had previously been identified by an unknown source as a platinum print. Note the neutral image hue and off-white paper highlights
National Gallery of Art, Washington, Alfred Stieglitz Collection 1949.3.380

OPPOSITE PAGE, RIGHT
12. Alfred Stieglitz, *Katharine N. Rhoades at '291'*, 1915/1916, platinum print with a trace of palladium. This photograph had previously been identified by an unknown source as a platinum print. The photograph, printed from the same negative as the image in fig. 11, has a very warm image hue
National Gallery of Art, Washington, Alfred Stieglitz Collection 1949.3.379

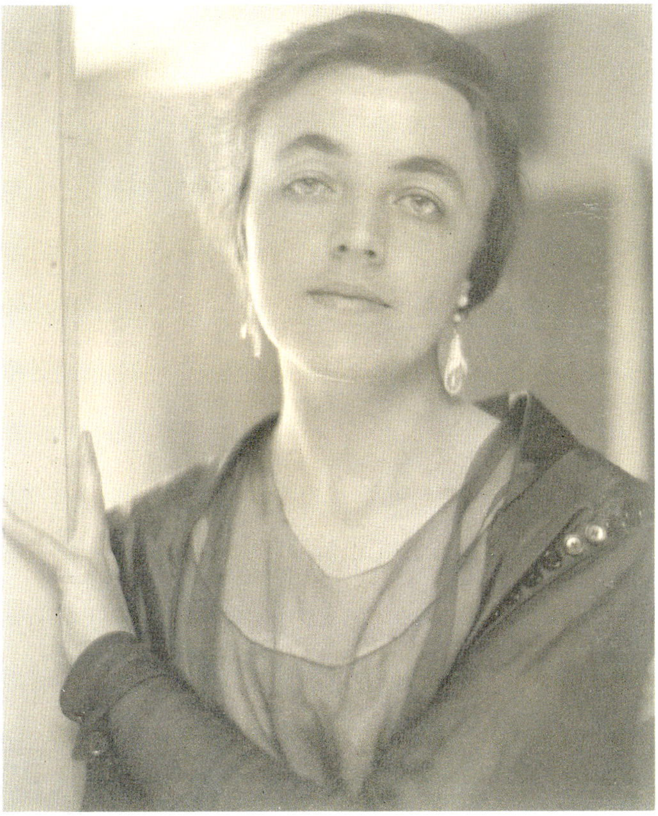

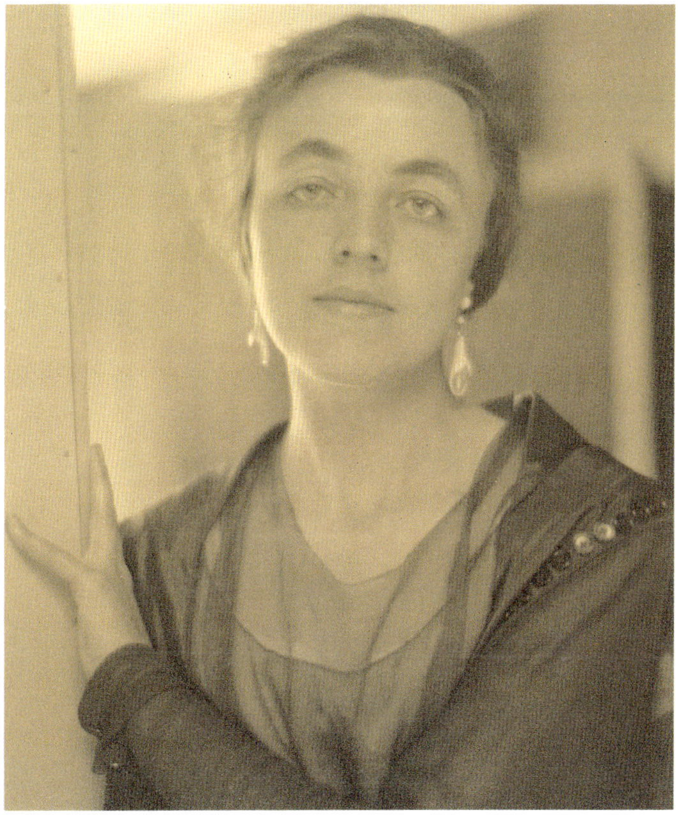

vertical scale=2,000

13. Detail, XRF spectrum of a dark area of fig. 11, with the background subtracted, using a barium chloride secondary target. The presence of platinum (Pt) is readily detected. Despite the neutral hue of this print, a trace of palladium (Pd) appears to be present. Compare this spectrum with that of fig. 3b, which is an XRF of fig. 1

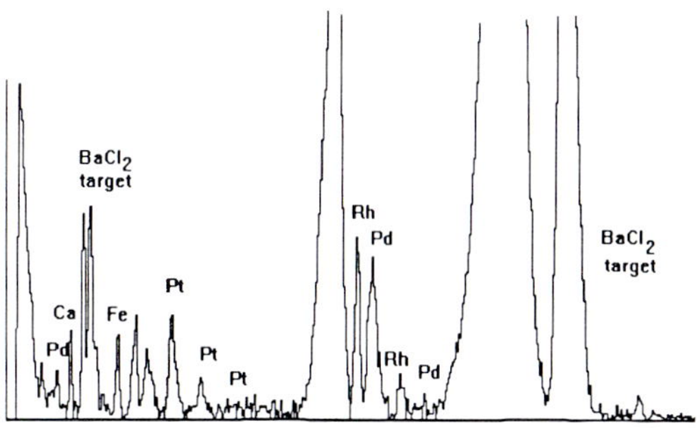

fig. 13 range: 40keV

vertical scale=2,000

14. Detail, XRF spectrum of a dark area of fig. 12, with the background subtracted, using a barium chloride secondary target. While analysis by XRF detected platinum as the primary image component, a trace of palladium (Pd) was also discernible

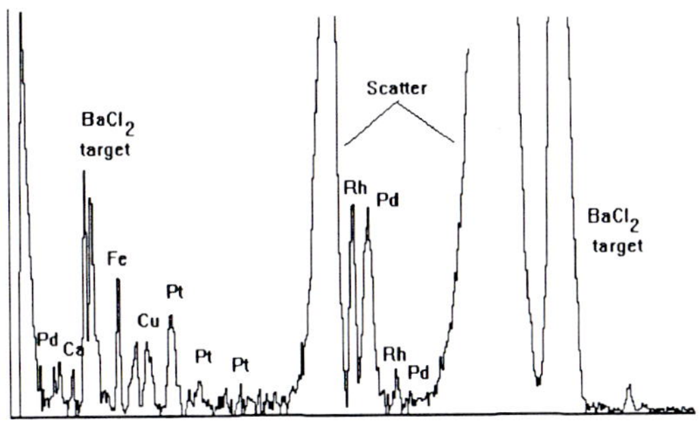

fig. 14 range: 40keV

A c. 1906 portrait of Stieglitz' daughter, *Kitty* (fig. 15), had been identified by an unknown source as a palladium print. For various reasons, including the date of the print and its unusually neutral image, this process attribution was questioned. Indeed, analysis by XRF revealed that the image consisted of platinum and mercury; no palladium was detected (fig. 16a–b). It is likely that this platinum print was processed with a mercury developer, which Stieglitz commonly used.

Another photograph, a portrait of Georgia O'Keeffe (not illustrated), had been identified by an unknown source as "black palladium or platinum." Analysis by XRF detected no palladium but confirmed the presence of only platinum. The photograph can now, therefore, be identified as a platinum print, thus

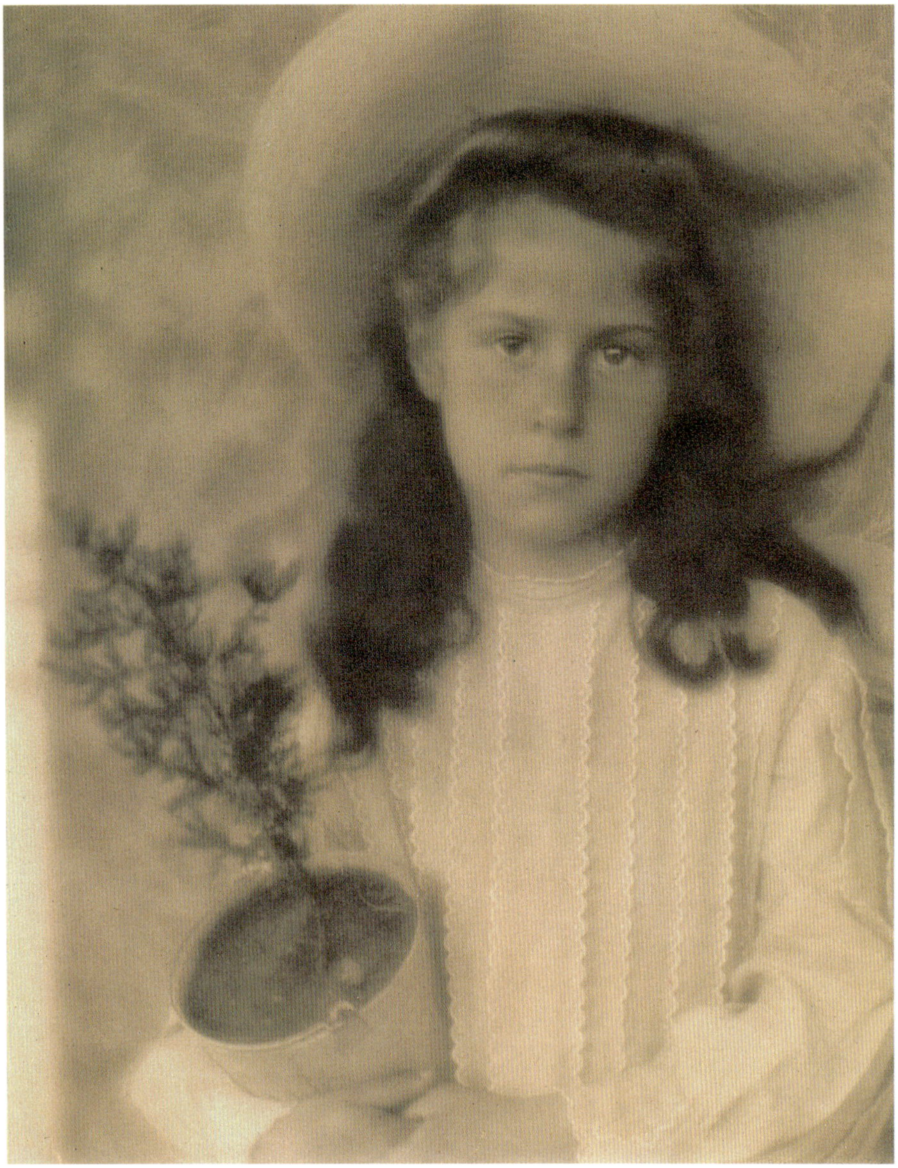

15. Alfred Stieglitz, *Kitty*, c. 1906, platinum print, probably processed with a mercury developer. This photograph had previously been identified by an unknown source as a palladium print. Analysis by XRF revealed the presence of platinum (Pt) and mercury (Hg), but no palladium

National Gallery of Art, Washington, Alfred Stieglitz Collection 1949.3.285

vertical scale=2,000

16a. Detail, XRF spectrum of a dark area of fig. 15, with the background subtracted, using a barium chloride secondary target. The presence of palladium is not detected in this print. Positive identification of mercury cannot be made with this target

vertical scale=1,000

16b. Detail, XRF spectrum of same, with the background subtracted, using a molybdenum secondary target. The platinum (Pt) and mercury (Hg) peaks are evidence of their presence. Compare this spectrum with that of the combination platinum-palladium–mercury-developed facsimile in fig. 7b. The trace amounts of mercury present in this print are at the edge of the detection limits of this instrumental protocol, but the fully resolved mercury peaks allow its positive identification

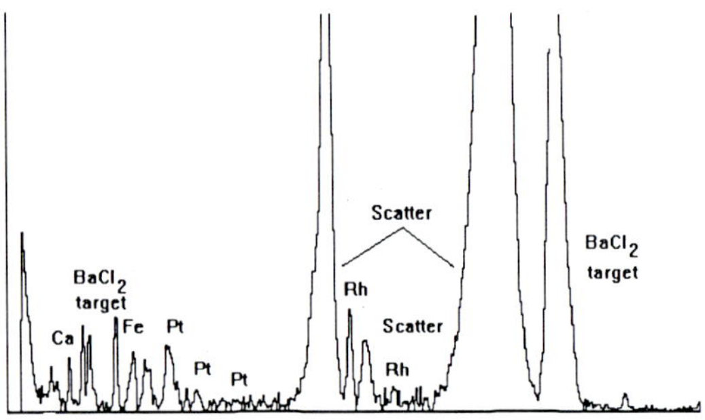

fig. 16a range: 40keV

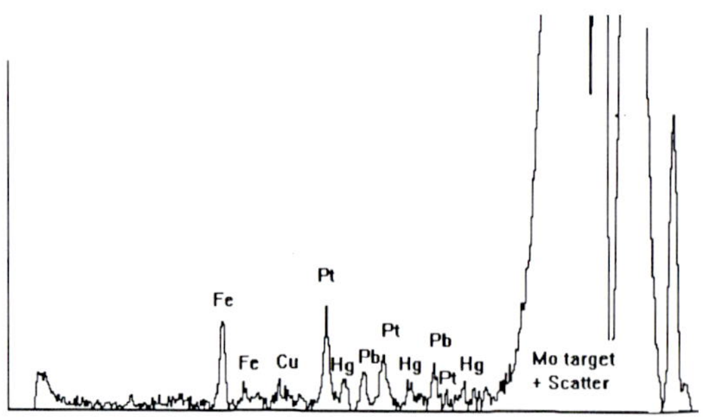

fig. 16b range: 20keV

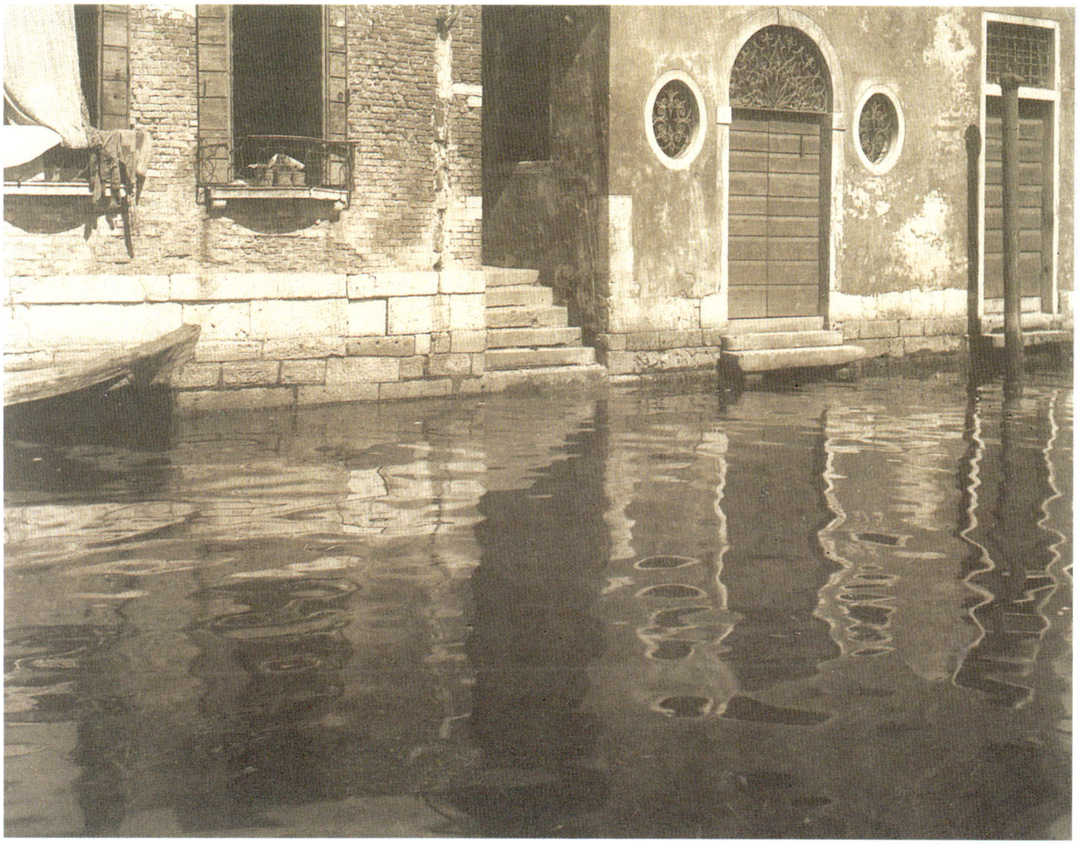

17. Alfred Stieglitz, *Reflections*, 1894, platinum print. The very warm image hue in this print is similar to that of fig. 1. Because only platinum (Pt) was detected, this photograph is believed to be an example of a warm-developed platinum print
National Gallery of Art, Washington, Alfred Stieglitz Collection 1949.3.136

vertical scale=2,000

18a. Detail, XRF spectrum of a dark area of fig. 17, with the background subtracted, using a barium chloride secondary target. The presence of platinum (Pt) is detected in this print

vertical scale=1,000

18b. Detail, XRF spectrum of same, with the background subtracted, using a molybdenum secondary target. It is interesting that, despite the similarity in image hue of this image and the mercury-developed *Self-Portrait, Cortina* (see fig. 1), no mercury or other toners were detected in *Reflections*

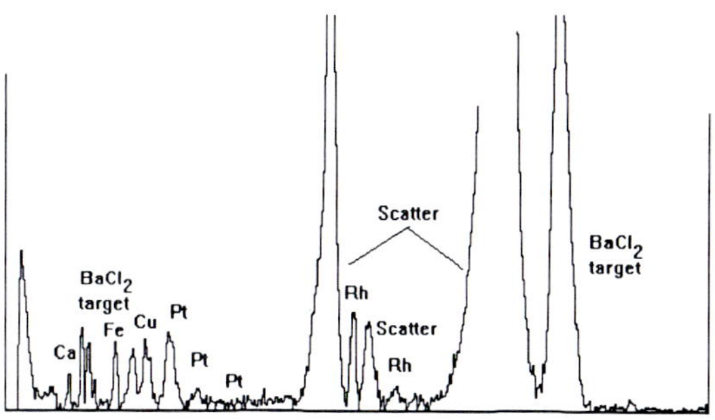

fig. 18a range: 40keV

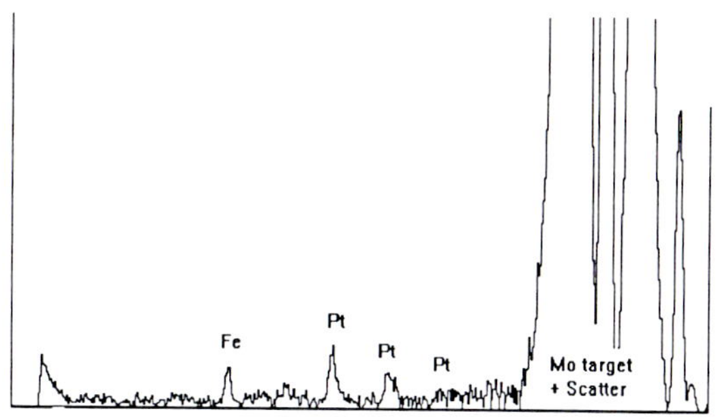

fig. 18b range: 20keV

clarifying the dubious process attribution.

Analysis using XRF of images with very similar warm hues, *Reflections* (fig. 17) and *Self-Portrait, Cortina* (see fig. 1), demonstrated a different point. The primary image material in both prints was found to consist of platinum. However, mercury was detected in *Self-Portrait, Cortina* but not in *Reflections* (compare fig. 3a–b to fig. 18a–b). While it is evident that *Reflections* is a simple platinum print, it is probable that *Self-Portrait, Cortina* is a mercury-developed platinum print. Perhaps the difference in appearance is related to the temperature of the developer. However, it is impossible to draw categorical conclusions regarding the exact proce-

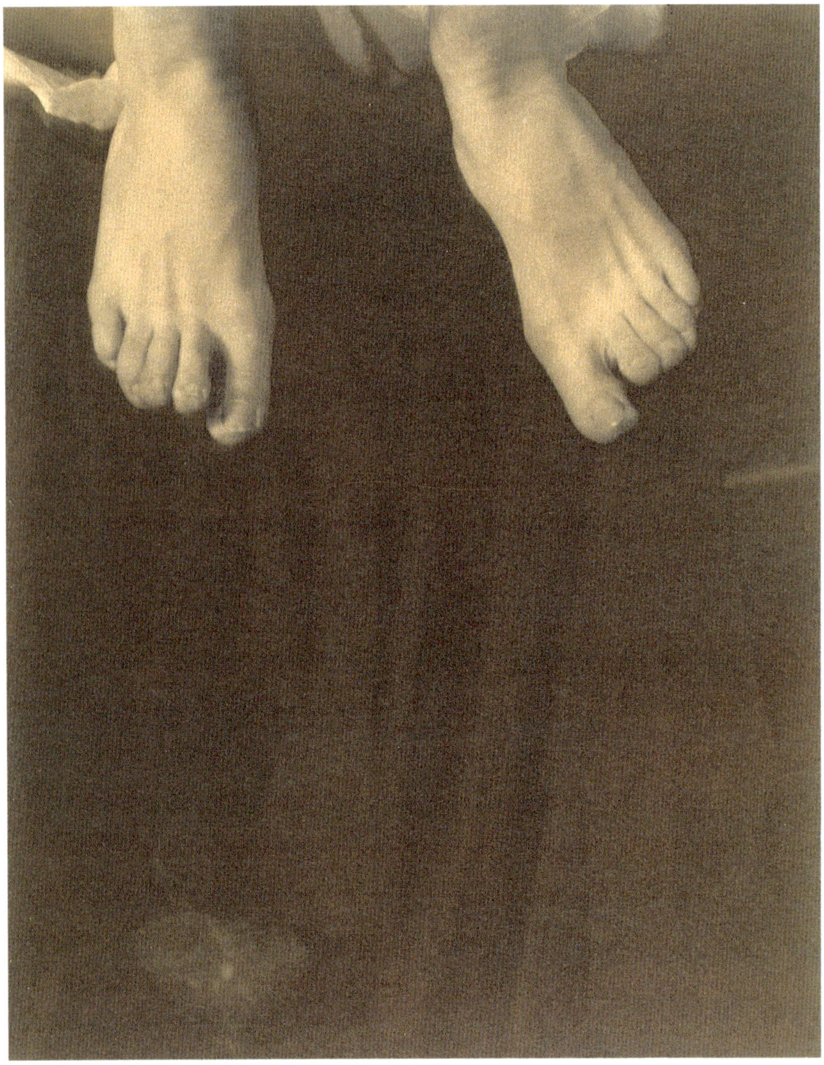

19. Alfred Stieglitz, *Georgia O'Keeffe: A Portrait—Feet*, 1918, palladium print, processed with a mercury developer. This palladium print has the appearance of many of Stieglitz' palladium prints, with its dark brown and brown-black image hues
National Gallery of Art, Washington, Alfred Stieglitz Collection 1980.70.54

vertical scale=2,000

20a. Detail, XRF spectrum of a dark area of fig. 19, with the background subtracted, using a barium chloride secondary target. While palladium (Pd) is detected using this target, mercury (Hg) is not detected. See fig. 20b for confirmation of the presence of mercury

vertical scale=1,000

20b. Detail, XRF spectrum of same, with the background subtracted, using a molybdenum secondary target, which clearly reveals the presence of mercury (Hg). The presence of palladium (Pd), however, cannot be confirmed with this target

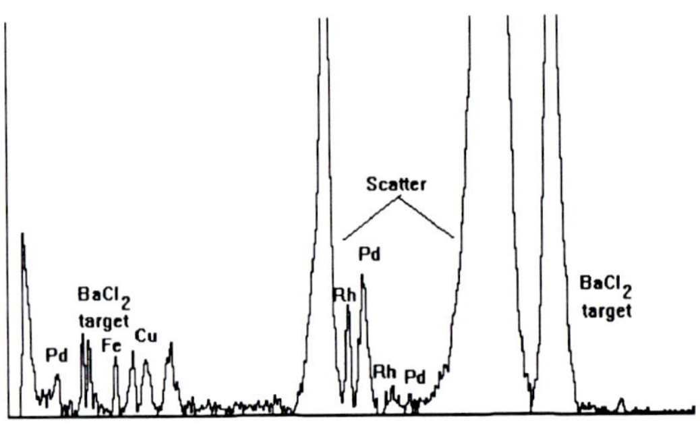

fig. 20a range: 40keV

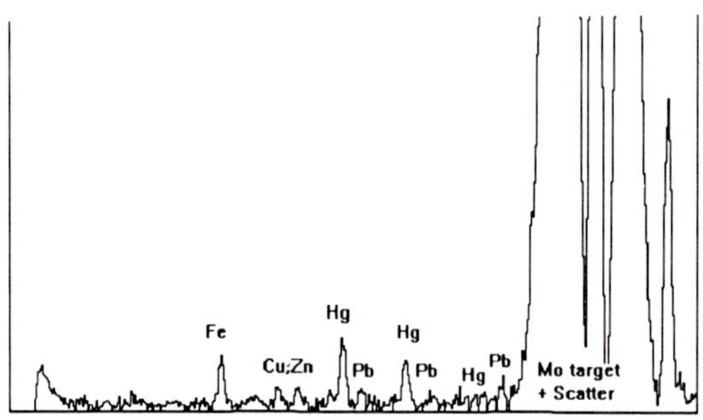

fig. 20b range: 20keV

dures to produce these very similar-hued prints. The two mercury-developed platinum prints (figs. 1 and 15) are strikingly dissimilar in appearance, another indication that a positive identification of process cannot be made by visual examination alone.

Another example of the usefulness of XRF analysis to establish process attribution is illustrated in *Georgia O'Keeffe: A Portrait—Feet* (fig. 19). The XRF spectrum shows the presence of palladium and mercury (fig. 20a–b). The presence of mercury indicates that this palladium print was probably processed in a mercury developer. This specific analysis is interesting in that, while it is often possible through visual examination to segregate Stieglitz' platinum prints from his palladium prints, it is impossible to ascertain whether mercury is present without the aid of chemical analysis.

Summary

As this study illustrates, the exact nature of photographs made by the platinum and palladium processes is difficult to determine by visual observation alone. Photograph conservators, historians, curators, and practitioners of the early processes with a trained eye may be able to narrow the possibilities concerning the process by which an individual print was made. The virtually unlimited variations of processes available to and used by photographers over the past 150 years, however, are manifested by a practically infinite assortment of photographic image appearances. The variety of subtleties and nuances displayed by platinum and palladium prints can be explained not only by their original exposure and processing but often by chemical and physical changes (such as fading, staining, and modification of surface characteristics) that have occurred over time.

Examination of the physical characteristics of photographic prints is essential to collections' care. XRF analyses can add important information regarding the chemical nature of photographs. The present study demonstrates that the materials that constitute the final image in a single-layered photograph can be confirmed using XRF. This analytical protocol requires that photographic standards are fabricated using known processes and are analyzed by XRF. Then XRF spectra of collection photographs made by unknown processes are compared to the standards' spectra. XRF is useful in reassessing previous assumptions about a specific photograph; analysis by XRF may uncover information that exposes commonly held assumptions as misleading or inaccurate.

The primary concern of this study was to evaluate the characteristics exhibited in platinum and palladium prints by Stieglitz. Therefore, facsimiles of other photographic processes were not fabricated for inclusion in this study. Further investigation by the authors into the use of XRF on other photographic processes will include the production of additional facsimiles. These facsimiles will include sample prints with a number of different ratios of platinum to palladium and the addition of compounds of lead and other elements to the facsimile photographs. In addition, the study will be expanded to include a variety of silver imaging systems with one or more component layers. All these variants of photographic processes influence the appearance and the preservation of photographic materials.

Conclusion

It is clear from this study that visual observation combined with XRF analysis will not necessarily reveal why a print has certain visual characteristics. Other information may be required (for example, known chemical method of production, previous conservation treatment, storage and exhibitions history) for a complete understanding of a print's appearance. While XRF is, in fact, useful in determining what metallic components are present in a photograph, it does not reveal the precise chemical structure of the final image materials or the manner by which a photograph was created. Other analyses, including transmission scanning electron microscopy, may reveal useful information such as metallic particle size. This information may help explain what makes a particular photograph appear the way it does.

NOTES

The authors wish to express their thanks to Julia Thompson, exhibitions assistant, and Sarah Greenough, curator, department of photographs; E. René de la Rie, head of the scientific research department; Shelley Fletcher, head of paper conservation; and Janice Gruver, editor for the conservation division, National Gallery of Art, Washington; Sarah Wagner, senior photograph conservator, National Archives and Records Administration; and Karen Garlick, acting head of conservation, National Museum of American History, Smithsonian Institution.

1. Alfred Stieglitz, "The Platinotype and the New York Amateur," *American Amateur Photographer* 4 (1892), 153–155; Stieglitz, "The Platinotype up to Date," parts 1 and 2, *Canadian Photographic Journal* 1, no. 11 (December 1892), 261–263; no. 12 (January 1893), 286–287; Stieglitz, "Uranium Toning of Platinotypes," *American Amateur Photographer* 5 (1893), 210; Stieglitz, "Sloppiness in the Platinum Process and Its Effect," *American Annual of Photography and Photographic Times Almanac for 1902* (1902), 28–30; Stieglitz and Joseph T. Keiley, "Platinotype in Colours: Controlled Development by Glycerine Process," *Amateur Photographer* 12 (4 May 1900), 351–355.

2. Sarah Greenough and Juan Hamilton, *Alfred Stieglitz: Photographs and Writings* (Washington, 1983), 230.

3. Douglas G. Severson, "The Effects of Exhibition of Photographs," *PictureScope* 32, no. 4 (Winter 1987), 133–135; Nancy Reinhold, "The Exhibition of an Early Photogenic Drawing by Henry Fox Talbot," in *Topics in Photographic Preservation*, vol. 5, American Institute for Conservation Photographic Materials Group (Washington, 1993), 89–94.

4. James L. Enyeart, Adelaide B. Anderson, Steven J. Perron, D. K. Rollins, and Quintus Fernando, "Non-Destructive Elemental Analysis of Photographic Paper and Emulsions by X-ray Fluorescence Spectroscopy," *History of Photography* 7 (April–June 1983), 99–113; Siegfried Rempel, "Qualitative Energy X-ray Fluorescence Examination of Historic Photographic Artifacts," paper presented at the Photographic Materials Group Winter Meeting of the American Institute for Conservation, Charleston, S.C., 22 March 1986; Valerie Baas, Judith J. Bischoff, and Leon Stodulski, "Ongoing Investigations into Chemical Enhancement of Faded Vintage Photographic Prints," in *Topics in Photographic Preservation*, vol. 5, addendum.

5. William Crawford, *The Keepers of Light: A History and Working Guide to Early Photographic Processes* (Dobbs Ferry, N.Y., 1979). See esp. part 2, "Ferric Processes: Platinum and Palladium," which describes, in detail, the procedures used to produce the facsimiles in this study.

6. Much has been published regarding the chemistry of platinum and palladium processes, yet the precise chemical nature of the reactions involved are not fully understood. See Adam S. Gottlieb, "Chemistry and Conservation of Platinum and Palladium Photographs" (B.A. thesis, Princeton University, Department of Chemistry, 1993). A mechanism for these chemical processes is proposed by Michael Ware, "An Investigation of Platinum and Palladium Printing," *Journal of Photographic Science* 34 (1986), 13–25, which should be consulted.

7. Stieglitz 1892, 154.

8. Bernard Edward Jones, *Cassell's Cyclopaedia of Photography* (London, 1911), 421.

9. Stieglitz 1902, 30.

10. Stieglitz 1892, 154, Stieglitz' italics.

11. For a discussion of process identification, see James M. Reilly, *Care and Identification of 19th-Century Photographic Prints* (Rochester, N.Y., 1986), 2–3.

12. Stieglitz, December 1892, 261–263, January 1893, 286–287.

13. Stieglitz 1893, 210.

14. W. De W. Abney and Lionel Clark, *Platinotype: Its Preparation and Manipulation* (New York, 1898).

15. Stieglitz to Paul Haviland, 22 April 1913; Stieglitz to George Eastman, 11 June and 1 December 1914; Stieglitz to R. Child Bayley, 9 October 1919; Stieglitz to Paul Strand, 6 October 1919, all in Alfred Stieglitz Archive, Yale University, New Haven, Conn., Collection of American Literature, Beinecke Rare Book and Manuscript Library, uncatalogued.

16. Stieglitz and Keiley 1900, 352.17. Stieglitz, December 1892, 262.

18. Crawford 1979.

19. Stieglitz, December 1892, 261–263 and January 1893, 286–287.

20. Stieglitz 1893, 210.

21. James McCorkle, "On the Uranium Toning of Platinum Paper," *Camera Notes* 6, no. 2 (1902), 90.

22. Eugene P. Bertin, *Principles and Practice of X-ray Spectrometric Analysis*, 2d ed. (New York, 1984), 21–49.

23. The lower limit of detection (LLD) has not yet been established for the metallic elements that may be contained in the final image material of a photograph. To determine the LLD, considerable extra work would have been required, and the results would have been applicable only to the XRF analyses performed with the setup and excitation conditions of the instrument at the National Gallery of Art. However, work is in progress to establish LLD in some practical form that would relate XRF detection limits to optical density in various types of photographs.

24. The elements that are of interest in this study can be detected using the freestanding instrument with the analytical protocol described in this paper.

25. The analysis was conducted under these conditions: The anode voltage and current were 60keV and 0.4mA, respectively, for the barium chloride target, and they were 40keV and 0.2mA, respectively, for the molybdenum target. Live accumulation time was 300 seconds. A Compaq PC 386 computer equipped with Kevex XRF Toolbox software provides for the processing and storage of the data.

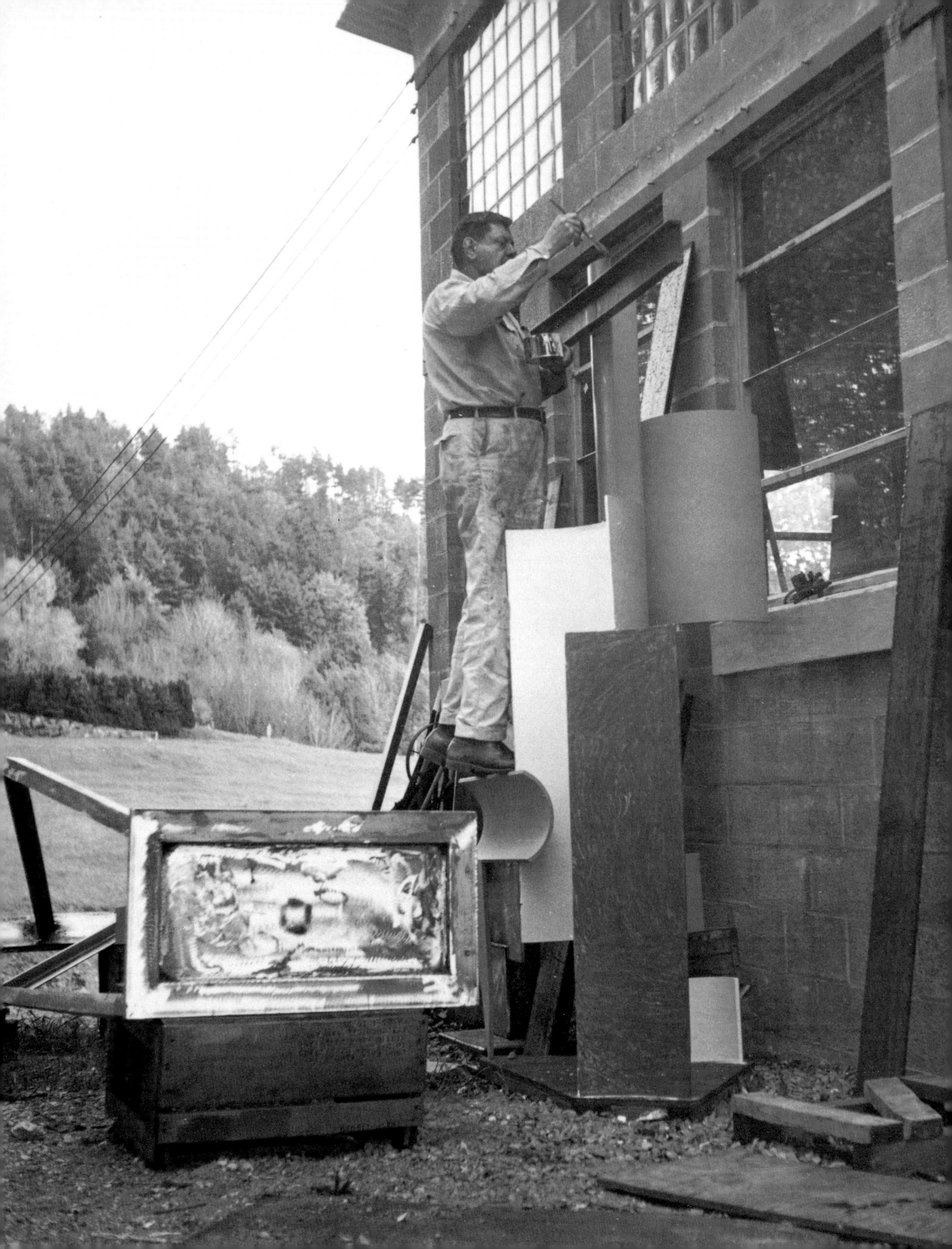

ALBERT MARSHALL

A Study of the Surfaces of David Smith's Sculpture

David Smith (1906–1965) created some of the most important three-dimensional art of this century. His construction techniques and their importance to contemporary sculptural practice and art history have been thoroughly explored.[1] Smith's surface-finishing techniques have been examined in a few publications and came to public attention in the 1970s during the controversy over removal of primer layers from several unfinished works by executors of Smith's estate and the poor condition of some sculpture at Smith's studio at Bolton Landing, New York.[2] Yet analysis of Smith's finishes—how they were used and how they have survived—has not been systematically undertaken.

Although the surface finish is the last process in the manufacture of the sculpture, it is the first to be lost over time. Throughout his career, David Smith was particularly interested in how his sculptures were finished. He lived during the century of steel.[3] His direct-metal technique relied on the use of cut and welded steel. The low-carbon steel he used until he could afford large quantities of stainless steel required a protective finish to prevent it from rusting if exhibited outdoors. As the protective coating on the sculptures became a part of the aesthetic mien of Smith's work, an investigation into his working methods and materials and their relationship to the aesthetic qualities of his work is important to conservators who are responsible for preserving the sculptures. David Smith's pronouncements about his finishing techniques for his sculpture proved to be a good starting point for a more thorough examination of his materials and techniques. Interviews with his first wife, Dorothy Dehner, as well as with his daughter Rebecca Smith and son-in-law Peter Stevens, helped to define a correlation between the artist's ideas and methods. Interviews with critic Clement Greenberg and study of Smith's notebooks from Bolton Landing and of those housed in the Archives of American Art, Smithsonian Institution, Washington, added insights. (See also the list of technical books in Smith's library at Bolton Landing, in the appendix.)

David Smith working on *Zig II*, 1961
Photograph by David Smith or photographer unknown, courtesy of Candida and Rebecca Smith

The Sculpture's Surface

Critical attention to sculpture in the mid-twentieth century shifted from the relationship of subject and form toward the formal play between surface and form. When cubist collages grew first into bas-reliefs and then into three-dimensional sculptures, their surfaces became more than the mere conveyors of form. In fact, surface became as important as the form, carrying the weight of aesthetic information. In essays such as "The New Sculpture" and "Modernist Sculpture: Its Pictorial Past," Clement Greenberg articulated his reductionist and optical theories in relation to sculpture. "Under the modernist 'reduction,'" he stated, "sculpture has turned out to be almost as exclusively visual in its essence as painting itself."[4] Greenberg applied his theories to the work of David Smith, and later they were elaborated by Rosalind Krauss. In *Terminal Iron Works*, Krauss explained that Smith's painted steel *Tank Totem IX*, 1960,

> *establishes itself as torso-become-surface. The surface seems legible as the whole object, since everything there is to know about the work is given to the eye in this reading. Everything has literally been made into surface, leaving no possibilities for the distinction between interior and exterior that had clung to Smith's earlier work, either because a quality of inside and outside had adhered to his still representational line or because of a sense of interior, no matter how vestigial, had remained to the sculpture as a whole because of its allusiveness. In achieving the surfaceness of* Tanktotem IX, *Smith achieves the surfaceness of painting, the conviction that everything in the work is coextensive with the surface, at last accessible to sight.*[5]

Smith had explained in 1950, "The difference between technical pursuits [in painting and sculpture] does not change the mind's reaction to form. Accent on any aesthetic difference is the prerogative of the layman. In my own case I don't know whether I make some pieces as painted sculpture or paintings in form."[6]

The Artist's Development

David Smith struggled with the concepts and forms of sculpture as he developed a direct-metal technique in the middle of this century. The industrial craft shop atmosphere permeated his working methods. He absorbed this atmosphere, along with basic skills, at the Studebaker automobile factory in South Bend, Indiana; the Terminal Iron Works in Brooklyn; and the American Locomotive Works in Schenectady, New York. The finishes with which Smith completed and hoped to preserve his sculpture were applied with the care and forethought of a modern industrial craftsman. Smith applied the same forethought to the crucial issue of what he believed the surface of sculpture should be. In an essay published in *Architectural Record* in 1940, he noted the lack of color in the sculpture of the recent past. Continuing in a polemic voice, Smith recited numerous modern industrial methods of coloring sculpture and described the modern industrial designer's use of color:

> *Contemporary sculpture has made timid use of color, although it has been an important factor in the best periods of the past. It is obvious that there exists a logic of color relationship to sculptural form just as there exists a logic in the scale of sculpture. Yet for centuries, bronzes have been dead dark, and marble, dead white. The public has acknowledged its preference for color in articles of everyday use, from hacksaw blades to automobiles. Ironically, the trend behind this change in "industrial design" is accountable to the fine arts, especially to cubism and later schools which regard texture, material, and color as aesthetic forces.*

It is of interest that Smith singled out cubism, essentially a painter's movement, in his address to architects. Pablo Picasso had painted bronze sculpture as early as 1914 (*Glass of Absinthe*); it was also Picasso and Julio Gonzalez, with their groundbreaking welded-steel sculpture in the 1930s, who inspired Smith to become a sculptor.[7]

The rest of Smith's *Architectural Record* article discusses methods of coloring and preserving metals, especially those intended for outdoor exposure. In addition to an interest in the purely visual at this early date in his career, Smith was also clearly concerned with the practical preservation aspects of finishing metal. In his own writing and in the critical writing about him it is evident that materials and techniques fascinated him from the beginning of his career. These concerns did not, however, divert him from his goal, as he always maintained that "the concept is primary, the medium secondary."[8] As he worked toward refining his concepts, he relied on many finishing techniques. In the early phase of his career Smith experimented with modern metal-finishing techniques. During the 1950s and into the 1960s he increased his output of sculpture both in quantity and in size; many of these sculptures had a painted finish. By the early 1960s he was able to afford stainless steel. In an interview with Thomas Hess, Smith said, "I like outdoor sculpture, and the most practical thing for outdoor sculpture is stainless steel."[9]

Technical Development

Smith began making his welded-steel sculpture at a small industrial welding shop, the Terminal Iron Works, where the solutions to technical problems were based on workability and expediency. The workaday atmosphere at the ironworks was very different from that of the atelier of a master artist, where students learned by following the rules promulgated in the established European system.

During World War II, Smith used his industrial skills as an employee of the American Locomotive Works. In his June 1942 letter of application to Sherman Miller in the design department, Smith described his previous experience:

> *My experience with various metals has covered the methods of fabricating, casting and forging with the use of finishing techniques such as metal spray, electroplating, fired vitreous enamels and oxidizing, depending upon the use, location, and specifications. I have also a working knowledge of paint technology.*[10]

The methods Smith used to finish some of his sculptures in the late 1930s—such as *Blue Construction* (fig. 1), enamel on steel, and *Structure of Arches*, 1939, "steel, with cadmium and copper plating"[11]—were probably the basis of the technical experience Smith cited in this letter of application. Smith's aesthetic foundations were therefore formed by the work of the cubists and surrealists, supplemented by his own art training, travel, and industrial experiences and observations.

Specific Techniques to 1950

As with most artists, Smith's early career can be characterized as a period of aesthetic and technical experimentation. His notebooks from these years are filled with technical concerns along with many formulas for

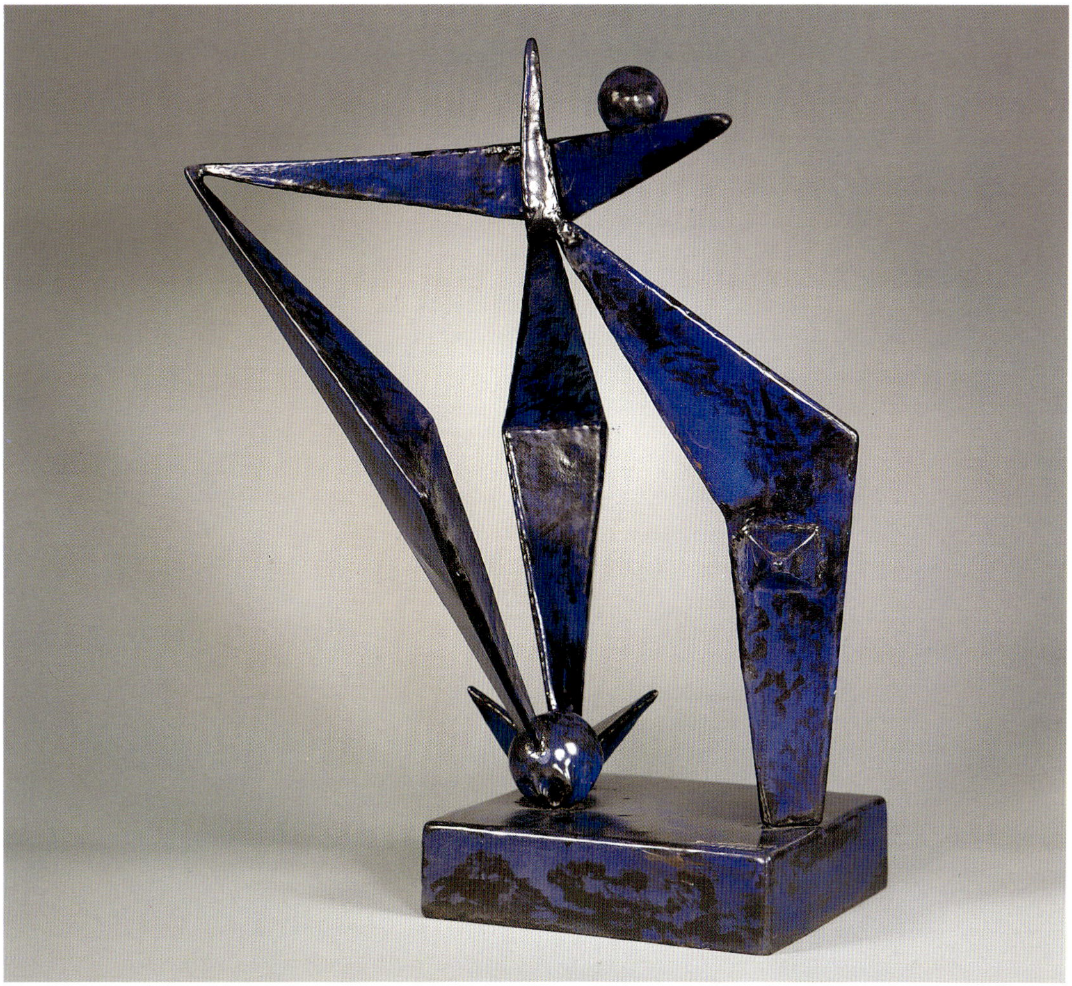

1. David Smith, *Blue Construction*, 1938, sheet steel with a baked-enamel finish
Collection of Candida and Rebecca Smith, on loan to the National Gallery of Art, Washington

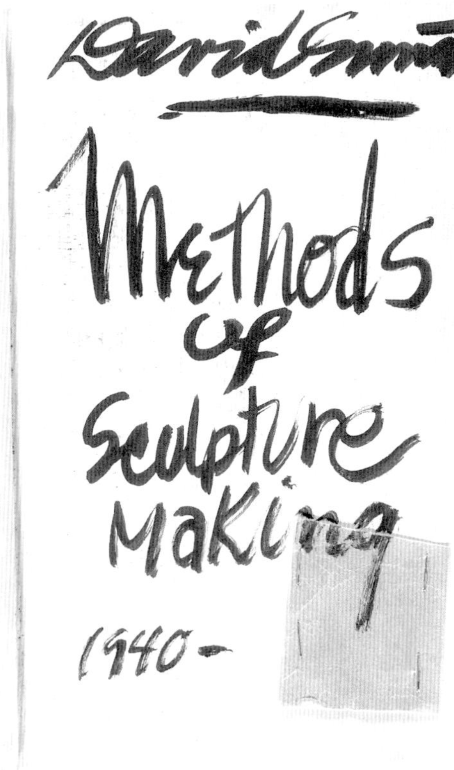

2. David Smith, notebook, 1940s, from his library at Bolton Landing, New York
Photograph by Lorene Emerson

patination solutions, descriptions of steelmaking techniques, and notes on various metals and metal-finishing techniques for his article in *Architectural Record*. Smith's notebook, "Methods of Sculpture Making: 1940–" (fig. 2), contains notes on casting methods, abrasives, and enameling, and a short history of bronze sculpture. The page titled "Enamel Sculpture" includes the following notes:

- enamels are vitreous or glassy substances—two kinds opaque and transparent—conditions to fulfill[:] preserve hard glassy appearance after fusion—must adhere firmly
- contract expand in same ratio with the metal to which they are applied
- insoluble—withstand acid & alkaline atmospheres—
- the base of metals is called fritz [*sic*] or flux over which are color layers containing various glassy substances of flint glass etc. colored with various fireproof metallic oxides.
- [a formula for] enamel cast iron [including] calcined flint, cornish stone, litharge, borax and potash. [The] powder mix fuse—when cold vitreous mass is ground sifted levigated (rubbed to powder with water—made into pulp [?] with water or gum water)
- smeared brushed—dried—fused in muffle. Refer jewelers notes.[12]

These excerpts reveal Smith's abiding interest in technically innovative approaches to the fabrication of his sculpture and his desire to explore the qualities of various surface-finishing techniques.

In his article for *Architectural Record* Smith described many of the new metal-finishing techniques that were developed in the 1930s:

The hard anodic coatings which can only be formed on aluminum are considered to be the most durable, have high corrosion resistance and high dielectric strength. They also produce the best surfaces for dyeing and mineral impigmentation. . . .

These copper-bearing alloys form oxide finishes from black through red, green, blue to brown by accelerated chemical action and can be fixed by special metal lacquers. To maintain a constant appearance, and for protection from stain to surrounding areas, bronze sculpture should always be protected by lacquer coats. . . .

Sculpture can be rustproofed by chemical immersion, or sprayed with a chemical solution to produce an insoluble black phosphate, and painted with stable oxide enamels. . . .

Steel sculpture sprayed with molten zinc or cadmium, which are both electronegative, thereby affording corrosion resistance by galvanic control, could then be sprayed with harder metals like chromium, Monel, silver, stainless alloys, cobalt, etc. By the molten-spray method, mixed metals can be applied, used as accents or as separate color areas. A steel sculpture sprayed with zinc and accented with copper would exhibit, after buffing, a silver-and-pink-colored granulated surface. This could be held by lacquer or permitted to take on a natural green and white-gray oxide patina. . . .

The important recent development is the "glass on steel" process of applying vitreous enamels to ingot iron sheets having a common coefficient of expansion. The firing of fabricated sculpture and fitted sections are both possible. So far, one sees the result of this process only on gasoline stations, hamburger stands, and stew pans, but the colors and freedom in application are as flexible and varied as those in the artist's oil palette.[13]

In 1952 Smith talked about his earliest work in a lecture to the New Sculpture Symposium, Museum of Modern Art, New York: "My work of 1934–1936 was often referred to as line sculpture, but to me it was as complete a statement about form and color as I could make. The majority of work in my first show at the East River Gallery in 1938 was painted. I do not recognize the limits where painting ends and sculpture begins."[14] A notebook entry titled "Gallery March," concerning this first exhibition, lists several sculptures and their finishes: *Agricola Head*, 1933, iron and steel painted red; *Aerial Construction* (fig. 3), iron painted a terra-cotta color; *Billiard Player Construction*, 1937, iron and encaustic; *Untitled*, 1937, iron painted orange, red-pink, white, black, brown, and blue.[15]

It is likely that *Aerial Construction* was made at the Terminal Iron Works. As the notebook entry states, the sculpture is iron, painted a terra-cotta color. Inspection of the sculpture in 1992 revealed that the finish is a dark red color scumbled over a black color. This scumbling technique was used frequently by Smith into the 1960s. An analysis of the paint medium by Christopher Maines, using pyrolysis gas chromatography–mass spectroscopy, revealed that the medium was paraffin wax.[16] A paint medium that is a mixture of wax (usually paraffin) and pigment applied in a hot (melted) state is called encaustic. Encaustic is a very durable medium; some of the oldest paintings now in existence are Egyptian Fayum portraits with a beeswax medium. The finish on Smith's *Aerial Construction* is in excellent condition.

Another sculpture from this period with a different finishing technique is *Blue Construction* (see fig. 1). Made of sheet steel (confirmed by x-radiography) with a baked-enamel finish,[17] it is an example of one of the industrial techniques Smith described in *Architectural Record*. Smith welded together steel plates to make the underlying form. A sample was taken of the surface material from a damaged area where the ball form meets the self-base. Examination of the sample under the microscope revealed that there is a thick, white layer of enamel over the steel substrate. Analysis of the enamel reveals that it is a soda-lime glass.[18] It is likely that Smith supervised the technicians

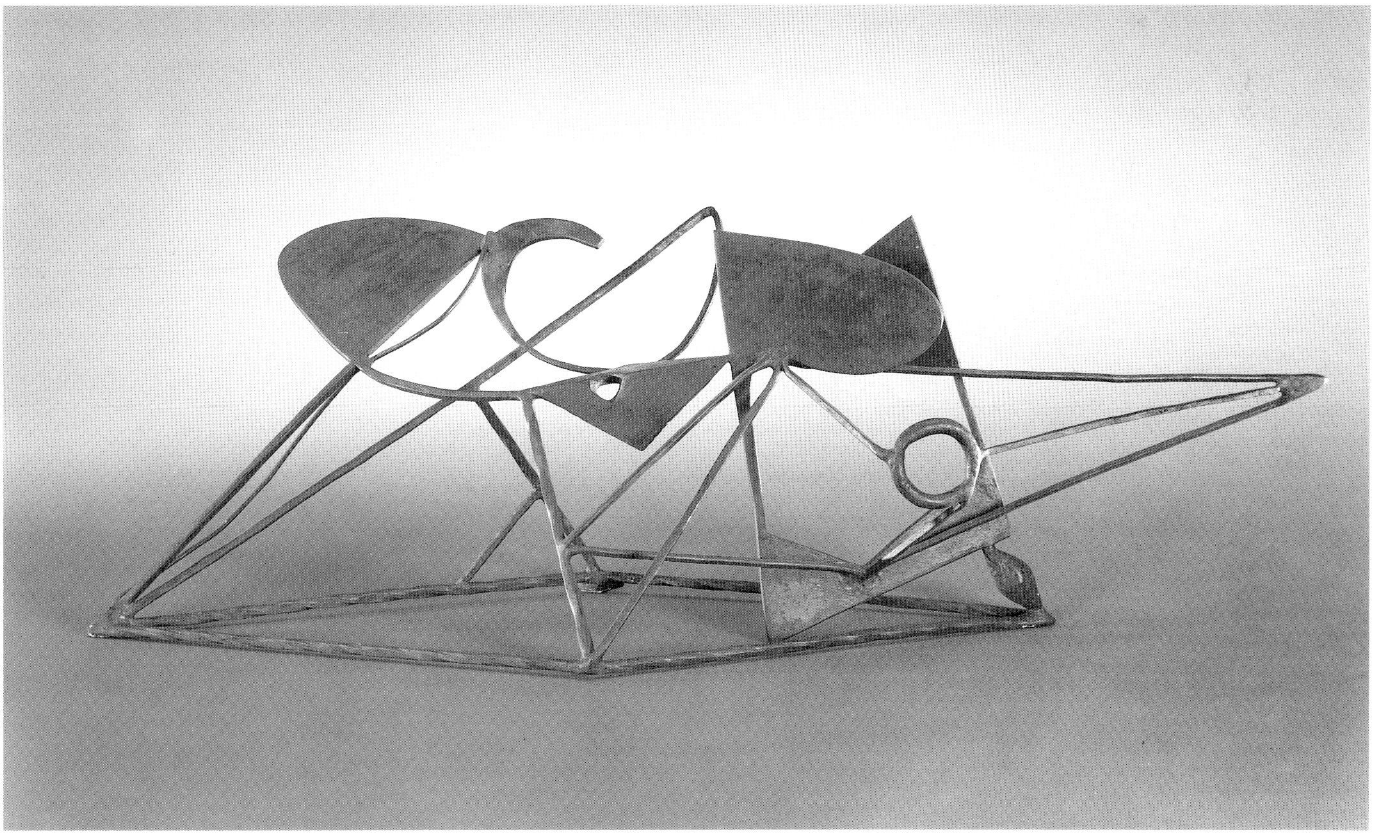

3. David Smith, *Aerial Construction*, 1936, painted iron
Hirshhorn Museum and Sculpture Garden, Smithsonian Institution, Washington, Gift of Joseph H. Hirshhorn

of an industrial enameling shop in the application of the fritted (ground to a powder) glass paste to the steel form and its subsequent firing, as he described in his notes on enamel sculpture. The surface has proven to be somewhat fragile. Due to the brittleness of the enamel, the edges have suffered many losses from small impacts. The flat areas of the surface, however, have remained in excellent condition.

Smith considered the surface of *Blue Construction* durable enough for outdoor exposure. In a journal entry he recorded "accepting responsibility for exterior use" in response to a letter from curator Dorothy Miller asking if *Blue Construction* could be displayed in the sculpture garden of the Museum of Modern Art, New York, "for awhile."[19] The variety of surface finishes employed on these early sculptures indicates that Smith did speak from experience in *Architectural Record*. Although he tried some of the techniques discussed, he did not use such experimental finishes for very long.

The technique of *Personage from Stove City*, 1946, anticipates the techniques Smith used in his mature work of the 1950s and 1960s.[20] This sculpture is welded together from stock pieces of steel (square in cross section, one-half in. on a side) and painted with several colors. A small cross-section sample was taken from the underside of the rocklike form in the "head," which is an unrestored area. The cross section showed a red primer layer beneath a black top layer. The binding medium proved to be drying oil,[21] typical for commercial paint in the 1940s. The sculpture had a thorough conservation treatment in 1982.[22] The original surface shows that the paint has an overall, even crack network typical of linseed oil paint. The paint has also darkened somewhat, again typical of a linseed oil paint. There is no indication that this sculpture has been exhibited outdoors. The paint layer and steel substrate are in good condition except for the typical cracking and darkening of the paint film.

Specific Techniques: The 1950s

In 1950 Smith received the first of two Guggenheim Fellowships. These grants enabled him to work full-time on sculpture for two years. Because welded-steel sculpture was still considered unusual at that time, Smith and Elaine de Kooning collaborated on an article to make it more understandable. "David Smith Makes a Sculpture" contains a detailed description of the materials and techniques he used at that time.[23]

The basis for the article was the creation of *Cathedral*, 1950, illustrated with accompanying in-progress photographs. De Kooning briefly described Smith and his working

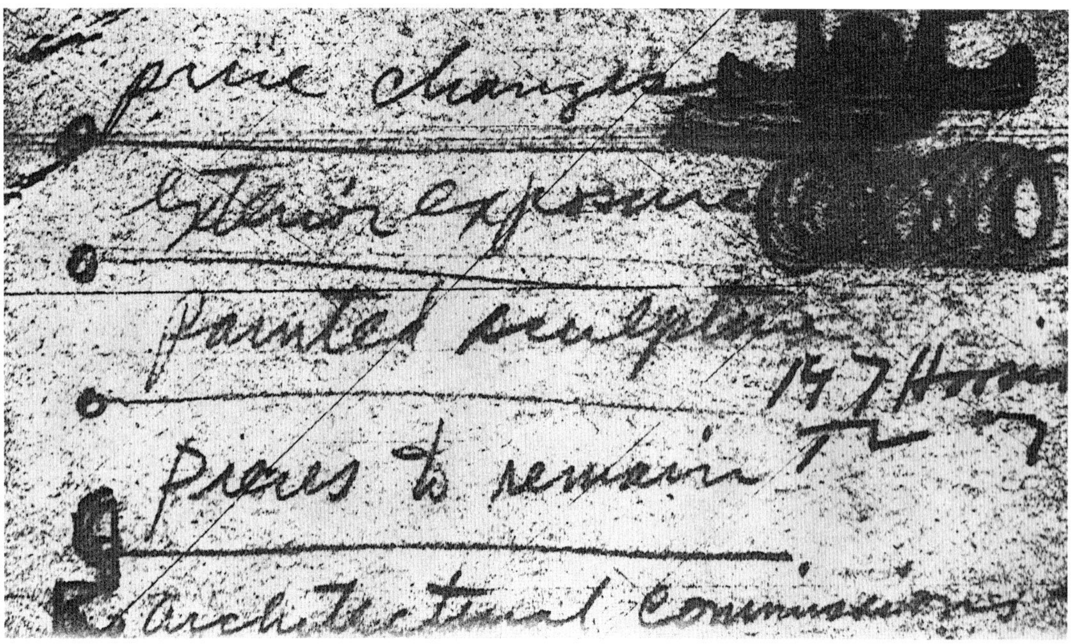

4. David Smith, notebook, early 1950s
Archives of American Art, Smithsonian Institution, Washington; author photograph

environment, as well as the intellectual and technical sources, for such unusual sculptures. The article culminates with details of how Smith finished *Cathedral* as well as other techniques he used at that time:

> *[Smith] rarely tries for a high polish, preferring a surface that expresses the crude nature of the metal. Using color in various ways this past year, he applied subdued metallic tones in even, mat coats, with a spray gun to some works, or smeared rust solvent mixed with large quantities of powdered pigment on others—like the huge, red* Fish *recently shown at the Whitney Museum—achieving brilliant, streaky, raw washes; but for the subtle, blond tones of* The Cathedral, *his method was more tentative. Dissolving splotches of rust, Smith coated the different metals of the pieces with a phosphoric acid, mixing small amounts of cadmium powder with it to produce deposits which varied from the golden patina on the steps to the mottled, whitish-pink of the twisted column, all falling into a unified range of shimmering, elusive tones.*[24]

Thirty-five years later, *Cathedral* was described as "steel, painted brown."[25] In the 1993 exhibition *Picasso and the Age of Iron* at the Solomon R. Guggenheim Museum, the "golden patina" and "mottled, whitish-pink" of *Cathedral* had, indeed, been replaced by a unified glossy brown.[26] The apparent need for conservation treatment, which resulted in the brown paint coating, is presaged in de Kooning's description of Smith's techniques. Smith evidently achieved a delicately beautiful surface, but without a meticulous preparation of the steel substrate and a careful buildup of the paint layers, the surface could not last for long, even in a relatively benign environment. What remains is Smith's sculptural form but not his surface.

The Guggenheim Fellowships made it possible for Smith to procure more materials and to increase greatly his output of sculpture. In 1951 and 1952 he completed more than forty sculptures. A page from Smith's early 1950s notebook reveals some of his concerns (fig. 4). "Price changes" indicates Smith's wish to keep his prices high. He was always arguing this point with his dealers and museums. One reason he charged high prices was that he had difficulty parting with his sculptures. As his inventory and the size of individual sculptures grew, so, too, did his storage problem. The solution was to store them outdoors. "I put my works in the fields. That was an emergency, lacking storage space. I did not conceive a field complex, but since it grew out of necessity, I accept it." "Exterior exposure" voices his concern that when steel is left outdoors it must be protected.[27] In his interview with Hess ten years later, Smith elaborated this concern:

> *I color them [my sculptures]. They are steel, so they have to be protected, so if you have to protect them with a paint coat, make it a color. Sometimes you deny the structure of steel. And sometimes you make it appear with all its force in whatever shape it is. No rules.*[28]

Paint was not always necessary, however. In a letter of 1956 to a collector who was concerned about leaving a sculpture outdoors, Smith wrote, "*Agricola IV*, owned by Mr. Arenberg, is shown outdoors on Arenberg's terrace. It is rusted, with a coat of protective wax. I like the color of rust and make much of my work for that finish."[29] Rust and steel are, of course, inexorably linked. While Smith liked the color of rust, he knew that the sculpture still needed protection. Unlike some sculptors whose emphasis is on the ephemeral aspect of a work's components, Smith did not consider the actual destruction of steel by rusting an essential part of his artwork.

At times Smith had difficulty deciding on the color of his sculptures, and he allowed many of them to remain for long periods unpainted or without the final top coat of paint. His notes described his method of storage, at least for smaller items: "I store it in the cellar warehouses of the house or envelop it in a protective navy specification wax (gun protection) and store [it] in one of the barns."[30] No receipts have been found, but his notes give a product description for "Cocoon Black #9—a black strippable protective coating,"[31] which can be applied before fabrication and was probably reputed to protect the steel until it was ready to be finished.

Agricola I, 1951–1952, is reportedly the first sculpture that was placed in the field, in 1954.[32] It continues to be displayed outdoors at the Hirshhorn Museum's Sculpture Garden and has probably been outside for most of its forty years. One would think that this sculpture would provide a good example of how Smith's painted sculpture has endured outdoors; however, it has been

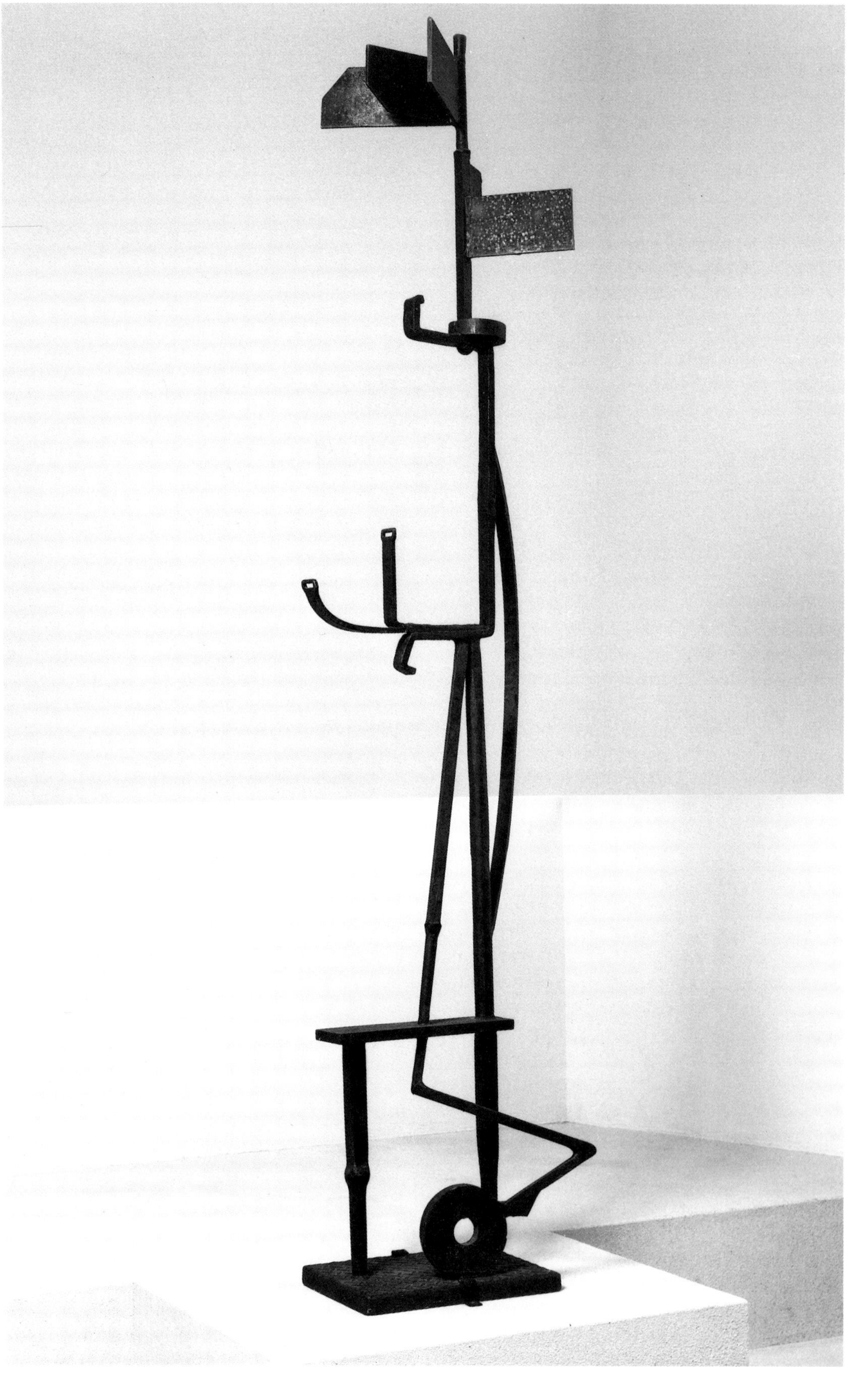

5. David Smith, *Sentinel I*, 1956, steel
National Gallery of Art, Washington, Gift of the Collectors Committee

6. **Cross section of paint from fig. 5, showing:**
1. rust
2. greenish primer over the steel
3. red lead oxide primer
4. black alkyd top coat

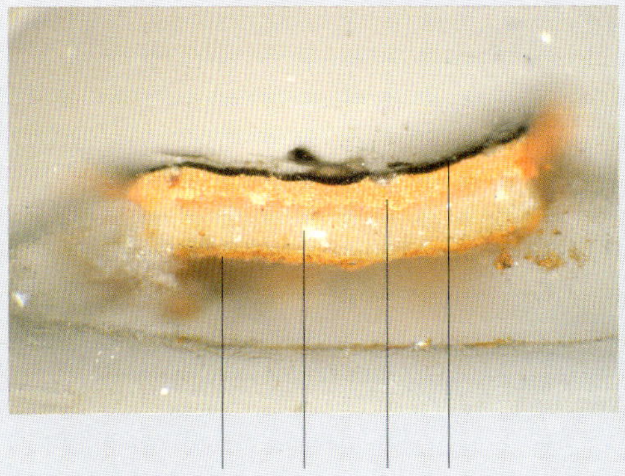

repainted many times over the years (perhaps even by Smith himself) whenever the paint surface began to show wear and fading. Conservation records at the Hirshhorn Museum indicate that in 1982 it was stripped of the buildup of old paint and repainted,[33] and it was again repainted in 1988 by conservators. An analysis of the 1988 paint indicates that the paint is an alkyd.[34] This paint had begun to chalk and crack by 1992, and *Agricola I* was repainted in 1994.[35]

Portrait of a Painter, 1954, is different from most of Smith's sculpture in that it is made of cast bronze.[36] Smith used cast bronze at various times throughout his career, beginning with the cast Medals for Dishonor series of 1939. *Portrait of a Painter* was made of cast bronze parts from found objects, including a printer's type box and an artist's palette. After welding the pieces together, Smith ground or wire-brushed the entire surface. The surface appears to be a combination of blue-green paint and chemical patina wiped away from the high points of the relief. The binding medium of the paint is an alkyd.[37] There appears to be a varnish on at least some parts of the surface. The condition of the surface is very good. While there are photographs of the sculpture outdoors, its surface seems to indicate that the work has not been outside for long periods. No conservation records have been found, but considering its present condition, the sculpture has probably undergone treatment in the past.

Sentinel I (fig. 5) is constructed of found steel pieces welded together to evoke a human sentry, as the title suggests, or perhaps a railroad semaphore. After Smith welded the pieces together, he probably followed his normal practice of grinding the welds and wire-brushing the surface. A cross section of the paint from *Sentinel I* (fig. 6), taken from the middle of the base, shows that Smith applied a double priming layer. The first or lowest layer (2) is a greenish primer over the steel (1). This layer resembles the green layer of *Zig V* (see fig. 8), analyzed by Christopher Maines. Maines' analysis indicates "the green primer layer contains no phthalates and hence is not an alkyd. Due to the high proportion of low-boiling point compounds, it is possible that it contains some large amount of a poly(vinyl)."[38] I believe, and Maines concurs, that this layer could be an "etch" primer, which according to Manfred Hess, an authority on industrial paints, is "based on a vinyl butyral resin, chromate pigment, and phosphoric acid."[39] The second or middle red-colored layer (3) is probably a lead oxide (known as red lead) primer, which was commonly used as a metal primer after about 1900. The top layer (4) is a black alkyd top coat.[40]

Sentinel I now exhibits some rusting through the paint layer and concomitant lifting paint. There is no record of this sculpture's having been outdoors for extended periods; however, it is noted in the files upon its acquisition by the National Gallery of Art in 1979 that "the piece has remained in the estate since the artist's death, and in storage at Bolton Landing, thus preserving its painted surface."[41] Stanley Marcus reported:

In 1957 Otto Gerson's Fine Arts Associates (FAA) purchased five sculptures, including Sentinel I, *for a substantial sum and planned a fall exhibition to open the week after the opening of the 'Mid-career' show at the Modern. . . . He [Smith] repurchased one sculpture from Gerson, as he did also from the Widdifield Gallery when he got into disagreements with them.*[42]

It is possible that this "one sculpture" might have been *Sentinel I*. In any case, no record of restoration has been found. Although the sculpture has been displayed indoors since 1979, the presence of rust below the paint layer suggests that the sculpture had been outdoors for a considerable period of time or that Smith did not clean off all the rust before he painted it. *Circle III* (see fig. 11), made six years subsequent to *Sentinel I*, has a similar primer structure, but its paint and substrate are in excellent condition. *Circle III* is made from off-the-shelf stock steel rather than found objects as was *Sentinel I*, and thus might not have had an inherent rust problem.

Specific Techniques: The 1960s

By 1960, Smith had reached a grandly mature phase in his career. Until his death in 1965 he finished his sculptures in three ways. The first, and most typical, was to build up paint layers over cleaned mild steel. The second, used most notably in the Voltri series, was phosphoric acid followed by a clear coating over cleaned mild steel. The third, used mostly in the Cubi series but seen as early as 1947 with *Aggressive Character*, was the use of stainless steel finished by finely scratching the steel surfaces with a carborundum grinding wheel.[43]

As discussed earlier, Smith usually found it necessary to provide a protective finish for the mild steel. Sometimes he used a non-pigmented drying-oil, alkyd, or varnish finish, as he did with most of the Voltri series. More often the steel was protected by paint, which involved using a color, and for Smith, color choice posed a persistent dilemma. In his discussion with Thomas Hess concerning his use of color he said, "I'm still working on that [color]. I've made two sculptures in tune properly between color and shape. But I've been painting sculpture all my life. As a matter of fact, the reason I became a sculptor is that I was first a painter."[44] As decisions on color often took Smith a long time, the time frame in which a sculpture was painted was extended. Lengthy intervals could, of course, affect the stability of the paint structure. If the various layers were allowed to become dirty between coatings, they could fail by delaminating. This sequence may be part of the explanation for the paint deterioration seen on *Rebecca Circle*, 1961, painted steel, in Dan Budnick's 1973 photographs in *Art in America*.[45]

In 1961, Frank O'Hara described the view of "the road up to the [Smith] house [as] a procession of new works, in various stages of painting."[46] In 1982, E. A. Carmean indicated that Smith studied the Circle pieces "for about a year before deciding on their coloration and returned to study them after they had been painted to see if changes were necessary."[47] Carmean cites Gene Baro's interview with Smith on the subject of painting his sculptures:

I don't have any wonderful ideas about painting. Sometimes, if I have made an error in mixing some paint some place, the wrong paint one place turns out to be the right paint that goes on this one, or else I pour a few cans together, and it comes out real nice, so you convince yourself that that's the right color for this one. I have to paint them wrong and look at them and sometimes change all the various old formalities, or even change the colors—it takes a little time, and it is funny where you draw the line. Sometimes you have to change the whole thing and sometimes it is kind of raw; and after you look at it a while, your laziness convinces you that you do not have to paint it again. I don't know where to draw the line. I guess you work on averages.

He continued on the subject of building up the surface:

The paint here is not artist's paint. It is auto enamel, and I use it; and it is much better than artist's paint for outdoors. First the iron is ground down so that it is raw, and it is primed with about fifteen coats of epoxy primer, and then a few coats of zinc . . . and then a few coats of white—and then the color is put on—after that; so it runs about twenty-five or thirty coats, and that's about three times the paint coat on a Mercedes or about thirty times the paint coat on a Ford or Chevrolet. And if it doesn't get scratched or hammered, I think the paint coat will last longer than I do. There is nothing better for outside painting than auto enamel, as far as I know.[48]

According to Marcus, Smith would build up a primer layer over raw steel and then paint the whole sculpture white.[49] Carmean noted that the sculptures then might remain in this state for months or even years out in the fields,[50] as can be seen in many photographs from the early 1960s of the fields at Bolton Landing. Smith would eventually work on the top coat, sometimes using a solid color, as in the Circle series, or create a complex painterly surface as in the Zig series.

In actually applying paint, Smith used whatever method would yield the results he envisioned. Even though Elaine de Kooning mentioned that Smith applied paint by spraying[51] and Smith used a spray technique in many of his canvas paintings during these years, Smith later said, in response to a question on the application of paint:

I usually brush because I don't like spray—it gets my skylights dirty, and it is bad for your health; and then all that spray goes all over everything, and the spray drips over all the other sculptures. I like brushing better, and when you use auto paint—that's made for spraying, actually—and sometimes you have to thin it a little more and put on more coats and thin it—but usually automobile paint goes on so smoothly that outside of a car it is not necessary to spray.[52]

Smith describes the second method of finishing mild steel in his "Report on Voltri." "*Voltri X* was painted with red lead, but all the rest [of the series] were cured with phosphoric acid, washed and lacquered," Smith recorded in 1962.[53] Phosphating is a simple and direct method of pretreatment of mild steel prior to painting. Its effectiveness is produced by the deposition of a "firmly adher[ed], chemically bonded layer of iron phosphate."[54] The "lacquer" described by Smith was probably an unpigmented drying oil or alkyd coating similar to Masury Oil or Val Oil. The executors of the Smith estate, Clement Greenberg, Robert Motherwell, and Ira Lowe, determined that most of these coatings had deteriorated by the late 1960s and, therefore, directed that they be restored with a similar coating.[55] *Voltri XVI*, for example, has an unpigmented, but yellowed alkyd coating[56] probably from the late 1960s. In describing the surface of the Voltri series sculptures, Greenberg said, "The absence in many cases of applied color or finish of any sort in these particular sculptures is part of the directness and part of the nervousness. The raw, discolored surfaces of the iron or steel members may be found a little repellent here and there, but by the same token they tend to efface themselves."[57]

The third method of finishing the surface did not involve a coating but rather the use of stainless steel as both a substrate and a presentation surface that did not require a coating to protect it from outdoor exposure. Smith had long known that stainless steel would be an optimum material for outdoor sculpture, and he had used it on a small-scale sculpture in 1947 for *Aggressive Character*. He started working on precursors to the Cubi series in 1956 with *Four Units Equal* and *Five Units Equal*.[58] In 1963 the series *Cubi I* through *Cubi XXVIII* was begun[59] and was still in progress at the time of Smith's death in 1965.

By the 1960s Smith no longer was experimenting with techniques such as the enamel on steel of *Blue Construction* (see fig. 1), which he had described in *Architectural Record*. The following discussion of four sculptures explains the three basic approaches Smith used to finish his sculptures. In his mature working methods he would either paint steel in single colors or in multiple colors with varied brushstrokes, clear-coat steel with a varnish or oil coating after phosphating, or use stainless steel finished by scratching with a grinding wheel.

Zig V (fig. 7) is an excellent example of an active, painterly paint surface on a welded steel form. The lower elements have a scumbled and partially wiped, black paint over blue; the upper elements are solid colors. A cross section of the paint layers of *Zig V* (fig. 8), taken from the right side of the reverse near the bottom edge, shows a double layer of a green primer (2) on top of the steel (1). The double layer is possibly an etch primer, "mostly based on a vinyl butyral resin, chromate pigment and phosphoric acid";[60] it contains some poly(vinyl) compounds.[61] The green primer is followed by a yellow zinc chromate layer in an alkyd binder (3).[62] There are then white (alkyd)[63] layers (4), which Smith consistently used to cover the sculp-

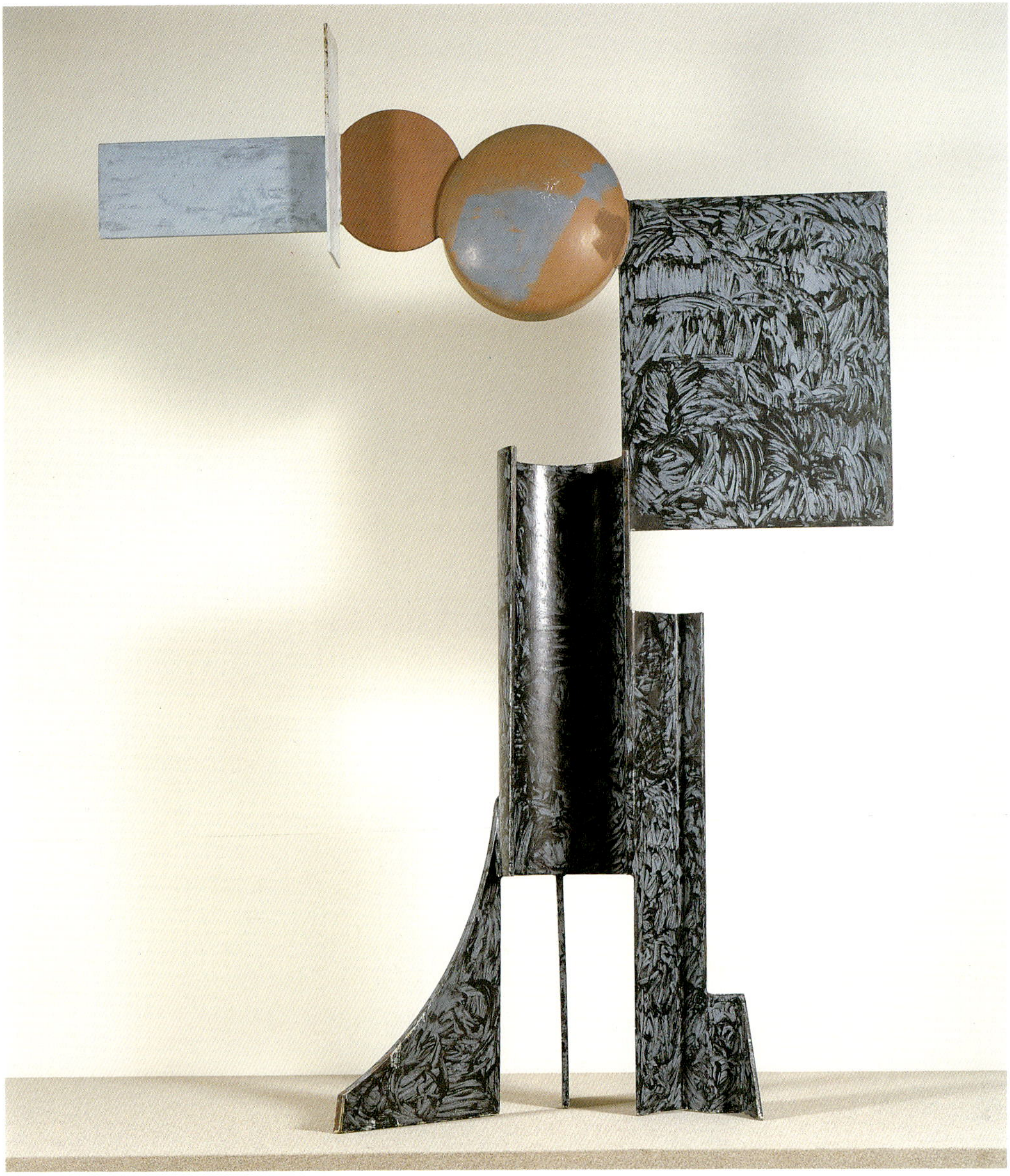

7. David Smith, *Zig V*, 1961, painted steel
Collection of Candida and Rebecca Smith, on loan to the National Gallery of Art, Washington

8. **Cross section of paint from fig. 7, showing:**

1. steel substrate with some rust
2. two coats of an unidentified green primer, which contains some poly(vinyl) compounds
3. several layers of zinc chromate in an alkyd binder primer
4. several layers of a white alkyd binder primer
5. blue alkyd binder decorative layer
6. black alkyd binder decorative layer, applied and partially wiped away

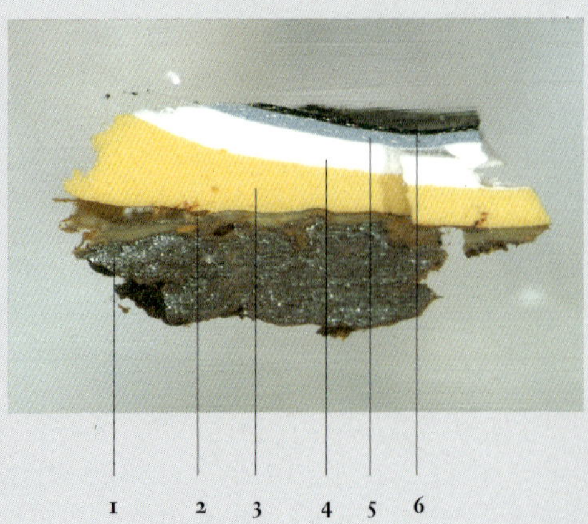

9. **Cross section of paint from fig. 10, showing:**
1. steel substrate with rust
2. green etch primer
3. several layers of yellow zinc chromate primer
4. white
5. red top coat
6. black top coat

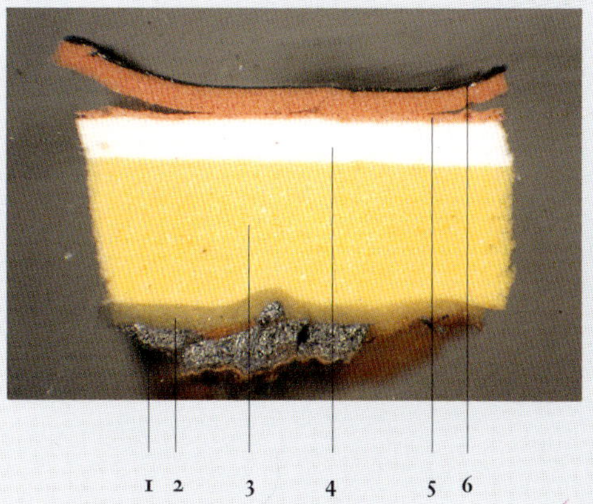

tures while he was deciding what to use for the final top coat. Finally, the blue top layer (5) with black (6) brushed over again, both alkyds,[64] can be seen. At present the surface of the sculpture displays chip losses along the edges and a few paint transfers. At least one chip loss on this sculpture was probably painted over by Smith. With a wide brush, the top center of the reverse edge of the blue and black area was painted over in the dull red color used for the larger sphere shape. A cross section (fig. 9) of the paint layers of *Zig III* (fig. 10) shows the same built-up paint system noted in *Zig V* (see fig. 8).

10. David Smith, *Zig III*, 1961, painted steel
Collection of Candida and Rebecca Smith, New York

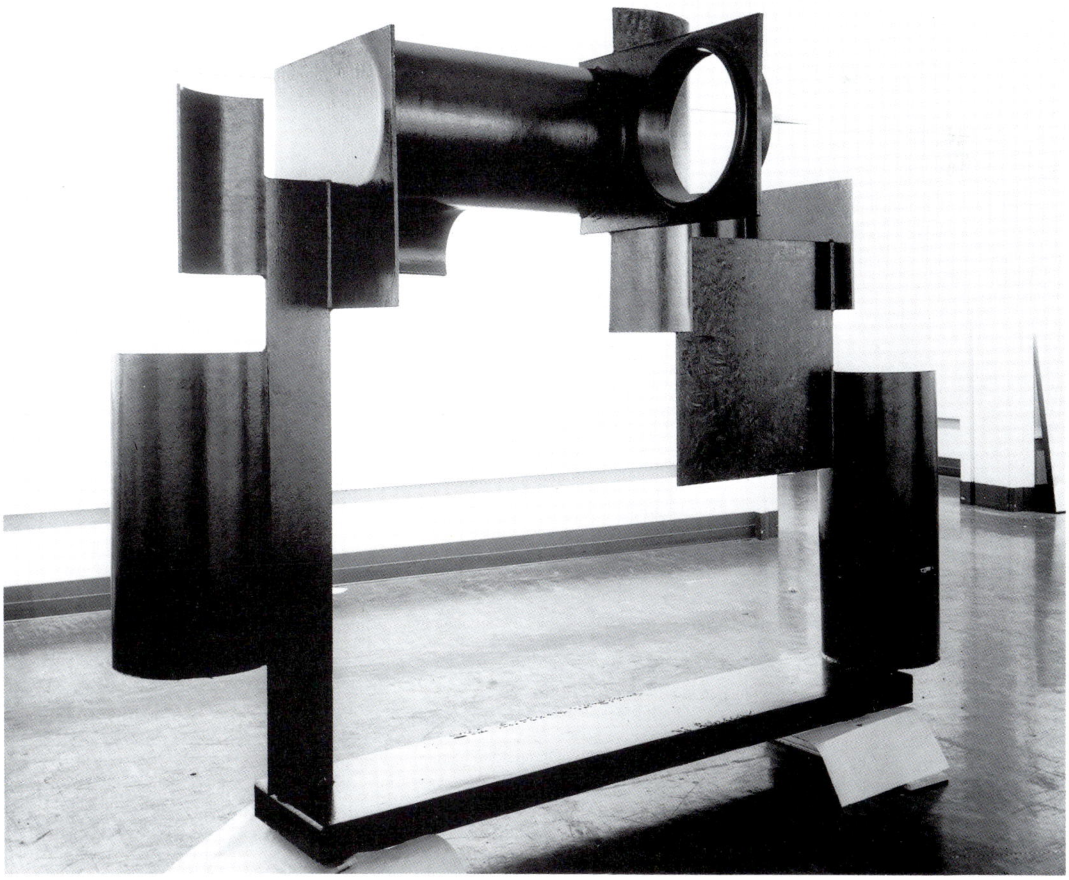

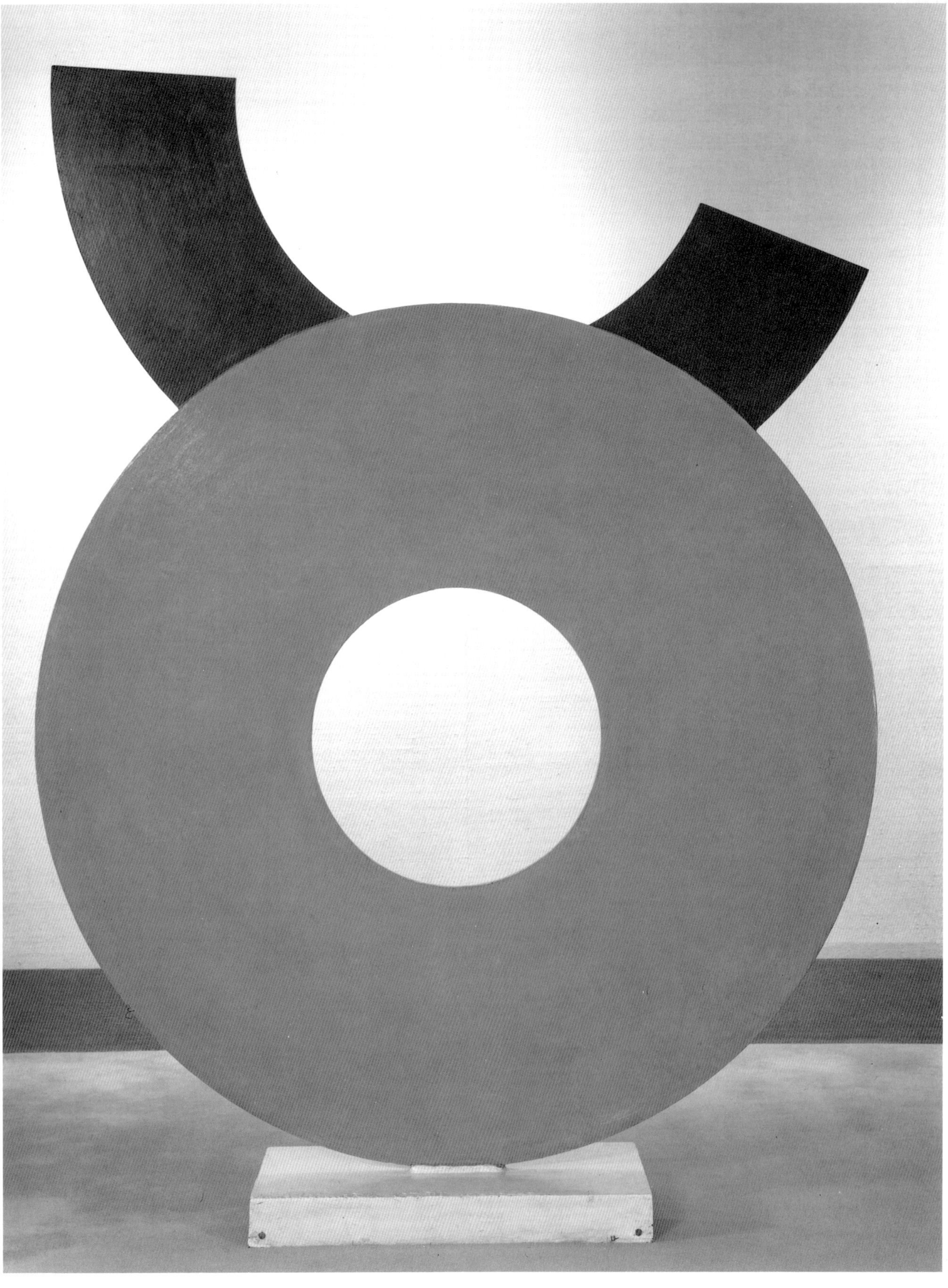

OPPOSITE PAGE
11. David Smith, *Circle III*, 1962, painted steel
National Gallery of Art, Washington, Ailsa Mellon Bruce Fund

Circle III (fig. 11) was constructed on 22 October 1962. It is made of one-half inch steel plate and bears a formal relation to Kenneth Noland's target paintings of the early 1960s. Photographs taken in the spring and summer of 1963 of the Circle series and other sculptures show them in a white state. It appears that *Circle III* was painted with a white primer and left in this state for approximately one year. A cross section of the paint layers (fig. 12) taken from the bottom front of the circle near the weld to the self-base shows the translucent greenish etch primer that contains some poly(vinyl) compounds[65] (1) over the steel. This primer is followed by a thin, red-brown primer layer of several coats (2). There is then a thick white primer (3) and a cadmium red top coat (4), both in an alkyd binder.[66] As mentioned, the condition of the paint on *Circle III* is excellent. *Circle I, II,* and *III* were sold in 1969 by Smith's estate to Marlborough Galleries, New York. Curatorial files of the National Gallery of Art note that "as a result of this early sale they [*Circle I, II,* and *III*] were among the very few painted works which are still in excellent condition, having been kept in the Marlborough warehouse rather than outside at Bolton Landing as were the other painted pieces."[67] Inside storage probably explains, in part, the condition of the surface. The rest of the explanation must be that the steel plate was from the mill rather than found objects, which were used for *Sentinel I*. Smith may also have cleaned the surface more thoroughly before applying the paint. *Circle I, II,* and *III* provide good examples of how carefully prepared surfaces with appropriate buildup of paint have survived for at least thirty years.

Voltri XVI is made of found objects of steel welded together at Voltri, Italy, for the Spoleto festival in 1962.[68] Smith said that all the Voltri series sculptures except one were treated with phosphoric acid and varnished. There is evidence on *Voltri XVI* of phosphorus and a coating of an unpigmented alkyd now somewhat yellowed from age.[69] This coating was restored in the late 1960s by the executors of Smith's estate.[70]

Cubi XXVI (fig. 13) was one of Smith's last sculptures. It is made of austenitic (chromium-nickel) stainless steel (T-304).[71] The sculpture requires no protective coating such as paint, varnish, or wax. Smith cleaned and visually unified the surface by finely scratching it with a carborundum wheel in random patterns. The surface problems associated with this finishing method, which Smith noticed early on,[72] are rust and other alteration products caused by the contamination of the carborundum wheels that had been previously used on other materials. This sculpture does not show the rust stains that Smith noticed with other sculptures. Perhaps by such a late date Smith knew to use a fresh, uncontaminated grinding wheel.

Smith's extraordinary sculptural accomplishments were made possible because his resources, both artistically and materially, were approaching an apogee. By the early 1960s he could afford large quantities of

12. Cross section of paint from fig. 11, showing:
1. unidentified green etch primer layer, which contains some poly(vinyl) compounds
2. several layers of a red-brown alkyd primer
3. several layers of a white alkyd binder primer
4. orange-pink (cadmium red) alkyd binder decorative layer

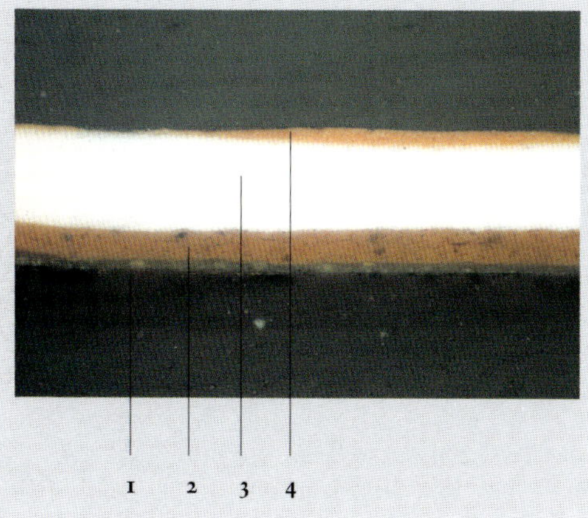

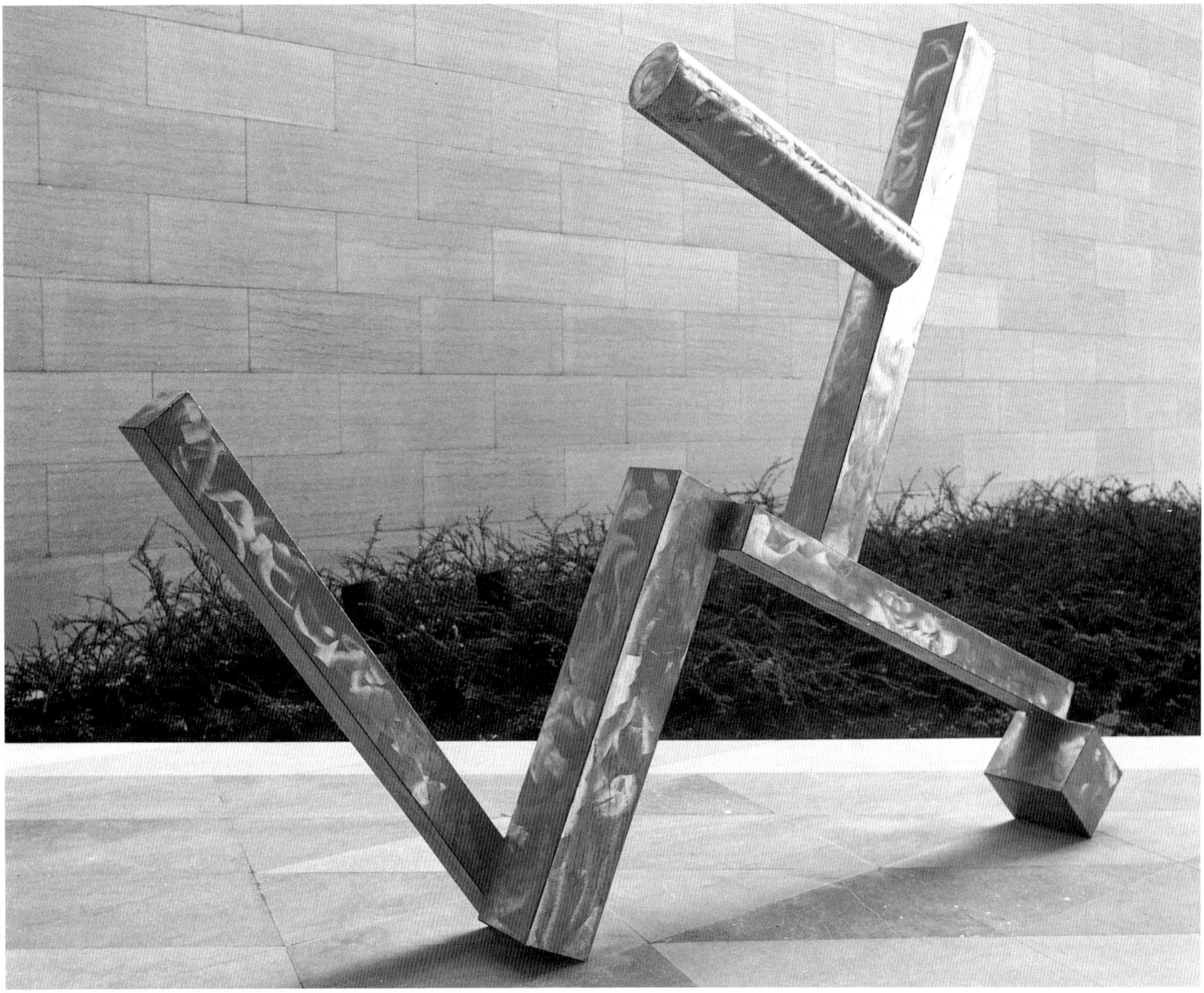

13. David Smith, *Cubi XXVI*, 1965, steel
National Gallery of Art, Washington, Ailsa Mellon Bruce Fund

stainless steel, and he could put it to grand artistic use. His sculpture's ground surface broke up the light that would hit it, achieving a brushstroke appearance similar to the effect in the Zig series of feathering different colors together on the surface. The immateriality achieved by the ground surface technique makes the sculpture approach the solely visual experience described by Greenberg and Krauss.[73]

The Artist's Method and the Industrial Method: A Synthesis

As we have seen, David Smith thought of himself as an artist who used industrial experiences to combine the practical aspects of welded steel with his own expressive vision. As steel is an important industrial material, considerable time and expertise have been concentrated on solving its biggest problem—rust. Steel was also an important material of David Smith's oeuvre, and he recognized that preservation of his sculptures required addressing the issue of rust. Where do industrial practice and the artist's choice converge?

To protect steel from rust, one or more of the following factors must be reduced or eliminated: an electrolyte such as water, ions, and a conductive pathway between the anodic and cathodic areas.[74] Sealing out the electrolyte and ions that promote electron mobility is the main function of coatings and paint. The factors favorable for fostering the best paint-to-steel seal are good adhesion to the steel substrate and low permeability of the paint film. Adhesion is promoted by an appropriate binder and a well-prepared surface. Low permeability of oxygen and various ions is promoted according to the type of binder used, the type and morphology of the pigments used, and the pigment-to-binder concentration.

Because there must be a good bond between the surface coating and the steel substrate, proper preparation of the steel surface is the first act of defense against corrosion. When a steel plate arrives from the vendor or mill, it "always carries a layer of millscale which, if completely intact and unbroken, would provide good protection to the steel."[75] The millscale has a laminated structure of different oxides of iron. Due to handling abrasions, the millscale is never intact. Manfred Hess reports:

> *Since millscale is cathodic to steel, such conditions are ideal for further rusting to ensue. Even after painting, moisture penetrating to the steel surface will cause electrochemical activity, the millscale will be undermined by rust formation, and it will eventually flake from the surface over large areas, bringing the paint film with it.*[76]

This electrochemical activity seems to explain the flaking problem shown in Dan Budnick's photograph of Smith's *Rebecca Circle*.[77] Marcus said that the method Smith used most frequently to preserve "the sculptures' surfaces from the elements before they were irreparably damaged . . . was to let the sculptures rust for a season, remove the [mill]scale with a wire brush, and then coat the surfaces with Masury oil."[78] Hess indicates that this is a common method of removing millscale, but he confounds the issue by providing information that "shows how unsatisfactory this procedure is."[79]

After cleaning the steel surface of millscale, the next step is a thorough degreasing and application of a corrosion-inhibiting primer. Sometimes the primer layer can be produced chemically by pickling, a procedure in which an acid is used further to clean and seal the surface. According to Hess, the best acid to use for this purpose is phosphoric because it dissolves most types of rust (but not millscale, interestingly enough)[80] and leaves a thin layer of iron phosphate to seal the steel surface from the elements. Smith was aware of this procedure and said he used the technique on the Voltri series.[81] Phosphating is also appropriate for painted steel; such a step is used in the painting of automobiles.[82] At present it is not known if Smith used the phosphating technique for his painted sculptures.

Smith used stock steel from the mill as well as found steel objects that had layers of rust in need of removal. In a 1965 interview with Gene Baro, Smith commented, "First [after the pieces have been welded together] the iron is ground down so that it is raw, and it is primed." Smith went on to say that he used auto enamel and described how he built up his paint layers in a way similar to the methods used on automobiles.[83]

After a phosphating step, automobile painters apply a corrosion-inhibiting primer. Optimally the primer should seal out moisture and oxygen and disrupt the flow of electrons between the anodic and cathodic points on the steel. D. A. Ansdell has written, "The most widely used [binding mediums for primers] in the industry were alkyd-based since they gave an acceptable balance between performance and cost, [and] epoxy types were introduced later to give improved corrosion resistance,"[84] and further that "currently, the most important barrier pigments are micaceous iron oxide and aluminum."[85] Pigments used in primers are either barrier or inhibitive types. Barrier pigments, by their morphology and orientation, help to reduce the penetration of water and oxygen through the paint film. Corrosion-inhibiting pigments include those based on lead, which "may inhibit corrosion by forming moieties such as soaps, which are adsorbed on the substrate,"[86] thereby helping to seal the surface, and those based on zinc, which works by a sacrificial cathodic mechanism (giving up electrons before the iron and thus corroding first).

Smith's description of his painting techniques closely follows the industrial guidelines given above. Smith said he used epoxy and zinc-chromate primers.[87] Marcus wrote that Smith used zinc-rich P-70 Pri-Met,[88] but evidence of Smith's use of epoxy primers on his sculptures has not yet been documented. At least four sculptures—*Sentinel I*, *Zig III*, *Zig V*, and *Circle III*—exhibit a translucent greenish primer layer over the steel that could be a vinyl butyral etch primer. This primer, like epoxy, can be in a two-part system;[89] therefore it is possible Smith thought of this primer as an epoxy. *Zig III* and *Zig V* have a yellow zinc chromate layer of about five coats, which is probably the P-70 Pri-Met described by Marcus.[90] There are a few coats of white paint beneath the surface (top) paint coat on *Zig III*, *Zig V*, and *Circle III*, as Smith had described to Baro. There is no evidence, however, that these four sculptures have the fifteen coats Smith described in the Baro interview.[91]

Observations and Summary

The properties of paint and their effect on the sculptural surface were central to David Smith's art. As he often said, he began as a painter and did not "know whether [he made] some pieces as painted sculpture or paintings in form."[92]

The evidence demonstrates that by the peak of his career Smith used the best available methods for coating steel. As a young artist he researched his materials, was employed in an industrial setting, and developed the ability to apply industrial craft to his artwork. In his formative years he used innovative techniques to finish some of his sculptures. *Blue Construction*, 1938 (see fig. 1), from this period, with its porcelain enamel coating, has been displayed outdoors at various times. The finish is in very good condition except where it has been chipped from improper handling, a function of the brittle nature of the enamel.

We have confirmed that Smith used encaustic on some early works such as *Aerial Construction*, 1936 (see fig. 3). This wax material is also delicate because it is very soft. There are photographs of this sculpture outdoors, but it was probably not kept outside for very long. The finish on *Aerial Construction* is also in very good condition.

Following the early period of experimentation with finishing materials, Smith moved to more traditional materials in the industrial lexicon, and after the early 1950s he developed an appropriate paint system, from pretreatment through primer to finish coat, that gave his mild steel works a good chance at survival in an outdoor environment. *Personage from Stove City*, 1946, is painted with a traditional linseed oil-bound paint. This work has probably not been displayed outdoors, and the paint film has survived well despite the inevitable craquelure of an aged, embrittled oil film.

From about 1950 on, it appears that Smith used paint with an alkyd binder. In common parlance this is our "oil" paint of today. At present it remains unclear if this alkyd paint was the "auto enamel" Smith said he used,[93] or if he used another industrial-maintenance type paint for his sculptures. The evidence seems to point toward the latter. Maines' analysis of the paint samples indicates that most of the alkyd paints Smith used seem to be in the long-to-medium oil length range.[94] These types of alkyds are often used in trade applications and architectural products, whereas the short-oil alkyds are used in industrial baking formulas for applications such as automobile paints.[95]

As we have seen, beginning with an example such as *Sentinel I*, 1956, Smith applied what was probably an etch primer whose properties might help to phosphate, bite into, and provide sacrificial corrosion-inhibiting pigment protection to the steel. The steel of *Sentinel I* has not fared as well as subsequent works we have investigated, such as *Zig III*, *Zig V*, and *Circle III*; I believe it has deteriorated because it was made from found materials that were not adequately cleaned before it was painted.[96] Smith appears to have improved on his paint system in the 1960s. *Zig III*, *Zig V*, and *Circle III* are not made from found objects and have additional priming layers of zinc chromate (*Zig III* and *V*) or red-brown (*Circle III*) and a buildup of white paint. The paint on these sculptures is in good condition except for some inevitable chipping caused by handling and moving of the large pieces as well as some abrasion losses (now restored) on the self-base of *Zig III*.

The phosphated and varnished finish of *Voltri XVI* has been restored, and its steel substrate appears to be in good condition. This type of finish benefits from being easy to monitor and easier to restore because of its clear coating. Smith's ultimate sculptures and his ultimate material—stainless steel—have, of course, fared very well over the past thirty years.

The study of David Smith's finishing techniques has provided the basis for our thinking and planning for the preservation of these important twentieth-century sculptures. Unless the extent of preparation of Smith's substrates is known for certain, it is not advisable to display the works outdoors for extended periods of time without close monitoring. Even if the sculptures are kept indoors, they should be monitored routinely. When conservation intervention is necessary, traditional point-by-point stabilization, filling of losses, inpainting, and coatings with appropriate acrylic polymers and/or waxes should suffice, with the necessary monitoring following treatment.

NOTES

The author would like to thank the Mellon Foundation for support during the research period of this project. At the National Gallery of Art, Washington, the support of Ross Merrill, chief of conservation, Shelley Sturman, head of object conservation, and E. René de la Rie, head of scientific research, were invaluable in the realization of the project. Christopher Maines, Mellon Fellow (1990–1993), scientific research department, spent long hours analyzing the paint samples and interpreting his findings, providing important keys to the research. Janice Gruver, editor for the conservation division, made many valuable suggestions for the document. Jeremy Strick and Nan Rosenthal, former curators, offered important suggestions and advice. The staff of the Archives of American Art, Washington, were extremely helpful during the hours of microfilm reading I did in the David Smith archives. Those knowledgeable about Smith and his work who granted me interviews were the late Clement Greenberg, Dorothy Dehner, and Rebecca Smith and her husband Peter Stevens, who were both extremely generous with their time and documents as well as in reading early drafts of the essay. Unless indicated otherwise, all photographs are reproduced with permission of Rebecca and Candida Smith.

1. Rosalind Krauss related many of Smith's technical accomplishments to his aesthetic achievements in *Terminal Iron Works* (Cambridge, Mass., 1971). E. A. Carmean's exhibition catalogue, *David Smith* [National Gallery of Art] (Washington, 1982), describes Smith's welding techniques and his painting of sculpture. Stanley Marcus describes the techniques Smith used throughout his career in *David Smith: The Sculptor and His Work* (Ithaca and London, 1983).

2. See Rosalind Krauss, with photographs by Dan Budnick, "Changing the Work of David Smith," *Art in America*, September/October 1974, 30–34, and subsequent letters to the editor, *Art in America*, March/April 1975, March/April 1978, May/June 1978.

3. Steel is iron with a low amount of carbon as an alloying element. Technically, steels with less than 0.15% carbon are classified as irons. George S. Brady and Henry R. Clauser, *Materials Handbook*, 11th ed. (New York, 1977). David Smith used the terms "iron" and "steel" interchangeably. Typically the steel he used for his sculpture—both the off-the-shelf stock sheet and found object pieces—was a low-carbon (0.15–0.30% carbon) mild steel, which he may have called "iron." Smith also made some use of cast iron, which is a brittle material containing more than 2% carbon (typically 3%).

4. Clement Greenberg, *Art and Culture: Critical Essays* (Boston, 1961), 139–145, 158–163, quotation on 142.

5. Krauss 1971, 98. *Tank Totem IX*, collection of Rebecca and Candida Smith, on loan to the National Gallery of Art, Washington.

6. David Smith, "What I Believe about the Teaching of Sculpture," address to the Midwestern University Art Conference, Louisville, Ky., 27 October 1950, in *David Smith*, ed. Garnett McCoy (New York, 1973), 64.

7. David Smith, "Sculpture: Art Forms in Architecture—New Techniques Affect Both," *Architectural Record* 88 (October 1940), in McCoy 1973, 44–48, quotation on 46. Pablo Picasso, *Glass of Absinthe*, Museum of Modern Art, New York.

8. David Smith, handwritten note [c. 1938], Archives of American Art (AAA), Smithsonian Institution, Washington, microfilm reel NDSmith 3, frame 482. The holdings of the David Smith archives are microfilmed notebook pages, loose pages, sketchbook pages, receipts, and other personal documents. Most handwritten notes are undated, but as they were microfilmed in the order in which they were found, suppositions can be made as to relative date. The microfilm reels are labeled NDSmith 1–5.

9. David Smith, "The Secret Letter," interview by Thomas B. Hess, June 1964, in *David Smith* [exh. cat., Marlborough-Gerson Gallery] (New York, October 1964), in McCoy 1973, 183–184.

10. Smith to Sherman Miller, design department, American Locomotive Works, 1 June 1942, carbon or draft copy of letter, AAA, microfilm reel NDSmith 1, frame 282.

11. Rosalind E. Krauss, *The Sculpture of David Smith: A Catalogue Raisonné* (New York and London, 1977), 23; *Structure of Arches*, collection of the Addison Gallery of American Art, Andover, Mass., gift of Mr. and Mrs. R. Crosby Kemper.

12. David Smith's notebook, titled "Methods of Sculpture Making: 1940–," was recently found at Bolton Landing by his daughter Rebecca Smith and son-in-law Peter Stevens. It remains in their possession.

13. Smith, "Sculpture," in McCoy 1973, 45–47.

14. David Smith,"The New Sculpture," in McCoy 1973, 82.

15. Smith, notebook [1930s], AAA, microfilm reel NDSmith 1, frame 077. *Agricola Head*, collection of Rebecca and Candida Smith, New York; *Billiard Player*, collection of Dr. and Mrs. Arthur E. Kahn. There are eight *Untitled* sculptures listed in Krauss 1977; this *Untitled* is probably the one in the collection of Frank Stella, New York.

16. Very small (typically less than 1 mm square from the top layer to the substrate to achieve a cross-section) samples were taken from several of David Smith's sculptures. Christopher Maines, Mellon Fellow (1990–1993), National Gallery of Art, conservation division, scientific research department, developed a technique for identifying different paint media (drying oil, alkyd, wax) using pyrolysis–gas chromotography–mass spectroscopy (gc–ms). We sampled several of David Smith's sculptures, as will be reported later in this essay. We analyzed an artist's oil paint dated from the 1940s, samples from a Mark Rothko painting (which served as a contemporary standard of an artist's oil paint), an automobile refinish paint dated from the 1950s (used by an artist, sample provided by the National Gallery of Art, Ottawa), and a sample of a commercial coating product, Val Oil. Simultaneous methylation–pyrolysis was used to examine all paint samples. A Chemical Data Systems (CDS) 120 pyroprobe with a platinum ribbon probe was connected to a Varian 3500 gas chromatograph with a 30 m x 0.32 mm DB-1 capillary column operated in split mode (100:1). The gas chromatograph was connected, via a fused silica transfer line, to a Finnigan MAT ion trap detector held at 250°C.

Alkyd paints are distinguished from pure drying-oil paints by the presence of a dimethyl phthalate peak at 920 seconds in the pyrogram. "Long" or "short" alkyds (used to distinguish the amount of drying-oil modifier: more for long, less for short) are distinguished by the proportion of the phthalate peaks to the various fatty acid peaks. The various layers of the paint samples were separated by the author under the microscope with a scalpel to obtain results for each paint layer. Christopher Maines, "Analysis Report," dated 12 November 1992, signed 4 and 6 January 1993, National Gallery of Art, Washington, conservation division, file (pysmith.rpt, 4 January 1993).

17. The finish is described by Krauss 1977, 12, as "fabricated steel, enameled blue, black."

18. Barbara Berrie, senior conservation scientist, National Gallery of Art, Washington, conservation division, scientific research department, personal communication.

19. Smith, journal [c. 1939], AAA, microfilm reel NDSmith 5, frame 003.

20. *Personage from Stove City*, collection of Rebecca and Candida Smith, on loan to the National Gallery of Art, Washington.

21. Maines 1993, 2.

22. This work was done by a private conservator in New York.

23. Elaine de Kooning, "David Smith Makes a Sculpture," *Art News* 50 (September 1951), 51.

24. De Kooning 1951, 51.

25. Jörn Merkert, ed., *David Smith: Sculpture and Drawings*, trans. Jörn Merkert and Hannelore Kersting [exh. cat., Kunstsammlung Nordrhein-Westfalen and Whitechapel Art Gallery] (Munich, 1986), 172, see also 69. *Cathedral* is in a private collection, New York.

26. Personal observation.

27. Smith, notebook [early 1950s], AAA, microfilm reel NDSmith 3, frame 1061; Smith, handwritten note [1960s], AAA, microfilm reel NDSmith 4, frame 704.

28. Smith, interview by Thomas Hess, in McCoy 1973, 156–163.

29. Smith to Harry Lockridge, 4 July 1956, AAA, microfilm reel NDSmith 2, frame 297.

30. Smith, handwritten note [1950s], AAA, microfilm reel NDSmith 4, frame 752.

31. Smith, printed product description [1950s], AAA, microfilm reel NDSmith 1, frame 510.

32. Carmean 1982, 56, *Agricola I*, collection of the Hirshhorn Museum and Sculpture Garden, Smithsonian Institution, Washington.

33. Lee Aks, conservator, Hirshhorn Museum and Sculpture Garden, Smithsonian Institution, Washington, personal communication, reported that he used the Derusto brand to paint the sculpture.

34. Maines 1993, 2.

35. Aks, personal communication. He used Derusto brand reddish brown matte-finish paint.

36. *Portrait of a Painter*, collection of Rebecca and Candida Smith, on loan to the National Gallery of Art, Washington.

37. Maines 1993, 3.

38. Maines 1993, 3.

39. Manfred Hess, *Hess's Paint Film Defects* (London, 1979), 239.

40. Maines 1993, 3.

41. Curatorial acquisition file for Smith, *Sentinel I* (1979.51.1), n.d., National Gallery of Art, Washington, department of twentieth-century art.

42. Marcus 1983, 131.

43. *Aggressive Character*, collection of Rebecca and Candida Smith, on loan to the National Gallery of Art, Washington.

44. Smith, interview by Thomas Hess, in McCoy 1973, 181–182.

45. Krauss 1974, 33. *Rebecca Circle*, Bolton Landing, New York.

46. Frank O'Hara, "David Smith: The Color of Steel," *Art News* 60 (December 1961), 32. The article is primarily devoted to Smith's painting of sculpture.

47. Carmean 1982, 49. The catalogue includes numerous photographs (see, for example, 58–59, 61) of the Circles and other sculptures in the fields in their white undercoats.

48. David Smith, "Some Late Words from David Smith," interview by Gene Baro, *Art International* 9, no. 7 (20 October 1965), 49–50.

49. Marcus 1983, 158.

50. Carmean 1982, 49.

51. De Kooning 1951, 51.

52. Smith, interview by Baro, 1965, 50.

53. David Smith, "Report on Voltri," 1962, in McCoy 1973, 163.

54. Manfred Hess 1979, 237.

55. Clement Greenberg, personal communication, 27 June 1991.

56. Maines 1993, 3.

57. Clement Greenberg, "David Smith's New Sculpture," *Art International* 8, no. 4 (May 1964), 36.

58. *Four Units Equal*, private collection; *Five Units Equal*, collection of the Storm King Art Center, Mountain View, N.Y.

59. Krauss 1977, 116–121.

60. See the discussion of *Sentinel I* above, and Manfred Hess 1979, 239.

61. Maines 1993, 3.

62. Maines 1993, 3.

63. Maines 1993, 3.

64. Maines 1993, 3.

65. Maines 1993, 3.

66. Maines 1993, 3.

67. Curatorial acquisition file and provenance file for Smith, *Cubi III* (1977.60.3), National Gallery of Art, Washington, department of twentieth-century art.

68. *Voltri XVI*, collection of Rebecca and Candida Smith, New York.

69. Glenn Gates, x-ray diffraction (XRD) for iron phosphate, in-house analysis report, 26 February 1993, David Smith conservation file, National Gallery of Art, Washington, scientific research department; Maines 1993, 3.

70. Greenberg (personal communication, 27 June 1991) could not recall the name of the product, but others have suggested, and I agree, that this coating was Masury Oil.

71. Marcus 1983, 162.

72. Smith, handwritten note [1960s], AAA, microfilm reel NDSmith 4, frame 935.

73. Greenberg 1961; Krauss 1971.

74. Ferrous metals rust. Iron, like other metals, tends to revert to its natural mineral state of ferric oxide. See J. A. Graystone, "Coatings for Buildings," in *Paint and Surface Coatings*, ed. R. Lambourne (Chichester, 1987), 418 and sec. 9.11: "Free energy considerations show that the reaction of iron with oxygen and water is energetically favoured."

D. A. Ansdell, "Automotive Paints," in Lambourne 1987, 437, gives a concise description of the process:

The steel surface, or substrate, is made of a lattice of iron, carbon dissolved in iron, iron carbides, and undissolved carbon particles. Rusting, or corrosion, is essentially an oxidation process aided by electrolytic action. On a moist steel surface electrolytic cells can be set up by the impurities in the metal (carbides, carbon, etc.) which give rise to cathodic (negative) and anodic (positive) areas. In the presence of water and salts (an electrolyte) these areas become the electrodes in a corrosion cell. Ferric ions form at the "cathode," where the metal becomes pitted, and combine with oxygen in the air and hydroxyl ions from the electrolyte to form iron oxide and hydroxide. Corrosion consequently starts to take place, and to prevent it the corrosion currents must be prevented from flowing by insulating the metal surface.

75. Manfred Hess 1979, 221.

76. Manfred Hess 1979, 221.

77. Krauss 1974, 33.

78. Marcus 1983, 157, based on Marcus' conversation with Clement Greenberg, 21 May 1978.

79. Manfred Hess 1979, 221.

80. Manfred Hess 1979, 225.

81. Smith, "Report on Voltri," in McCoy 1973, 163.

82. Ansdell 1987, 435.

83. Smith, interview by Baro, 1965, 49.

84. Ansdell 1987, 438. Ralph A. Williams, *Automotive Finishes*, Federation Series on Coatings Technology, Unit 25 (Philadelphia, 1981), 9, notes that water-thinnable epoxy ester vehicles [media] were used on the production line in 1972.

85. Ansdell 1987, 421.

86. Ansdell 1987, 421.

87. Smith, interview by Baro, 1965, 49.

88. Marcus 1983, 158.

89. Manfred Hess 1979, 239.

90. Marcus 1983, 158.

91. Smith, interview by Baro, 1965, 49–50.

92. Smith, "What I Believe about the Teaching of Sculpture," in McCoy 1973, 64.

93. Smith, interview by Baro, 1965, 49.

94. Maines 1993, 4. Oil length in alkyds refers to the proportion of synthetic or natural oil such as soya oil used to modify the alcohol-dibasic acid resin of the alkyd.

95. James R. Blegen, *Alkyd Resins*, Federation of

Societies for Paint Technology, Unit 5 (Philadelphia, 1974), 10.

96. Manfred Hess 1979, 211, points out:

Steel that has rusted during exposure to the atmosphere nearly always carries deposits of ferrous sulphate, formed in the presence of moisture and oxygen from sulphur dioxide in the air and iron. Ferrous sulphate crystals are very difficult to remove by scraping or wire brushing a rusty surface and if left behind are an important source of the breakdown of painted steel surfaces. The hygroscopic crystals promote rust under the paint film, give rise to filiform corrosion and since one volume of ferrous sulphate expands to approximately three volumes of rust, blistering of the paint film is inevitable.

This reaction could be what we see on *Sentinel I*.

APPENDIX

This list of technical books in David Smith's library at Bolton Landing was compiled by Rebecca Smith and Peter Stevens. A review of these titles offers insight into the information Smith sought while developing his working methods.

Alexander, W., and A. Street. *Metals in the Service of Man*. Middlesex, 1944.

Church, Arthur Herbert. *The Chemistry of Paints and Painting*. London, 1915.

Coffingnier, Ch. *Colours and Varnishes*. London, 1925.

De Wild, A. Martin. *The Scientific Examination of Pictures*. London, 1929.

Friend, J. Newton. *The Chemistry of Linseed Oil*. Chemical Monographs. London, 1917.

Gettens, Rutherford J., and George T. Stout. *Painting Materials*. New York, 1942.

Gott, Lester W. "Precision Casting of Low Alloy Steels." *Iron Age* (19 July 1945). Article found in Toch, *Paint, Paintings and Restoration.*

Heaton, Noël. *Outlines of Paint Technology*. London, 1928.

Herrington, Christiana J., trans. *The Book of the Art of Cennino Cennini*. London, 1899.

Hurst, George H. *Painters' Colours, Oils, and Varnishes: A Practical Manual*. London, 1901.

Laurie, A. P. *The Materials of the Painter's Craft in Europe and Egypt*. The Arts and Crafts of the Nations. London and Edinburgh, 1910.

———. *The Painters' Methods and Materials*. London, 1930.

———. *The Pigments and Mediums of the Old Masters*. London, 1914.

MacDonell, Ann, trans. *The Memoirs of Benvenuto Cellini*. N.p., n.d.

Merrifield, Mary Philadelphia. *The Art of Fresco Painting as Practiced by the Old Italians and Spanish Masters, With a Preliminary Inquiry into the Nature of the Colours Used in Fresco Painting, With Observations and Notes*. London, 1952.

Rigaud, John Francis, trans. *A Treatise on Painting by Leonardo da Vinci*. London, 1877.

Toch, Maximillian. *Materials for Permanent Painting*. New York, 1911.

———. "New Aids for the Detection of Picture Defects." *New York Times Magazine*, 25 August 1935. Article found in Gettens and Stout, *Painting Materials*.

———. *Paint, Paintings and Restoration*. New York, 1931. Inscribed: "To Dorothy and Smith with the regards of Maximilian Toch."

———. *The Protection and Decoration of Concrete*. New York, 1931.

CATALOGUE
DES
ESTAMPES
GRAVÉES
D'APRÈS
RUBENS.

Auquel on a joint l'Oeuvre de JORDAENS, & celle de VISSCHER.

Avec un secret pour blanchir les Estampes & en ôter les taches d'huile.

Par R. HECQUET, Graveur.

A PARIS,

Chez BRIASSON, Libraire, rue S. Jacques, à la Science.
CHARLES-ANTOINE JOMBERT, Libraire du Roi pour l'Artillerie & le Génie, rue Daupine, à l'Image N. D.

M. DCC. LI.

Avec Approbation & Privilége du Roi.

MARK STEVENSON

Print Restoration in Northern Europe: Development, Traditions, and Literature from the Late Renaissance to the 1930s

Print restoration is an integral component of the history of print collecting and dealing. Its legacy continues to affect the appearance of prints and the profession of art conservation. An exploration of the development of print restoration affords a greater understanding of the world of old master prints and the traditions of restoration practiced by printmakers, collectors, historians, and dealers. Developments in print connoisseurship are often, for example, reflected in their treatments. Establishing when a dealer or collector undertook or was willing to pay a restorer to create, improve, or replace an element of a print can help determine when that element became important.

The purpose of this paper is to trace the history of the restoration of intaglio prints, define the traditions of practice, examine the literature produced and relied on by those who restored prints, and identify those who made significant contributions to the field. The choice of the terminal period is based on the gradual changes in approach and attitude toward the remedial treatment of works of art that occurred during the 1930s and gave impetus to the developing profession of art conservation. The introduction of synthetic materials in restoration treatments, new analytical techniques, government-sponsored research into methods of restoration, and changes in the print market brought on by the worldwide depression of the 1930s mark the decade as transitional.

1. Robert Hecquet, *Catalogue of Engraved Prints after Rubens. To Which Are Joined the Work of Jordaens and Visscher. With a Secret for Bleaching Prints and Removing Spots of Oil* (Paris, 1751)
Library of Congress, Washington, Prints and Photographs Division, Case Book Collection

In this paper, a "print restorer" will be understood as any individual who attempted to mend, bleach, improve the appearance of, or physically stabilize a print. The designation includes collectors, dealers, bookbinders, printsellers, picture framers, housewives, and artists in addition to those who pursued print restoration as a trade. Describing a practitioner as a "print restorer" does not certify a grasp of aesthetic concerns, a level of manual skills, or an adherence to ethical standards. Still, within the broad group of those who restored prints, traditions of practice developed both in applications and in the literature. Differentiating these traditions will shed light on the motivations and aptitudes of print restorers.

The Status of Prints in the Sixteenth Century

Broadly speaking, in the sixteenth century loose-sheet prints served basically three functions as inexpensive substitutes for paintings, models and instructive material for artists, and supplementary illustrations. Even though instructions for the restoration of paper were published in the late sixteenth century, the functions and status of prints suggest that few were restored. Owners tended to value visual content over physical condition and were generally willing to substitute copies for originals. Presentation methods for prints consisted of tacking or pasting the impression directly to a wall or mounting the print on stretched fabric or a wooden panel and framing it. An application of varnish served to protect the print from dust and insect droppings and to make it appear more like the painting for which it was the surrogate. These presentation methods literally ensured the eventual destruction of the print. Such impermanence scarcely could have escaped the owner's attention, but, as the print had been chosen because it was inexpensive, the matter appears to have been of little concern. The same disregard for preservation is evident in the practices of artists and their apprentices. Adhering prints to mounts may have offered some protection, but frequent handling during drawing instruction was harmful. Pricking, tracing, and squaring—practices used to transfer images from prints—hastened their destruction or devaluation and are a further indication that in this period prints had not acquired status as works of art.

A collector's manual concerned with forming a *Kunstkammer* provided the earliest known description of an ideal print collection. In the *Kunstkammer*, prints served with other forms of graphic art as representational supplements to the curious, exotic, and rare objects assembled to demonstrate the endeavors of humankind and the marvels of nature. The guidelines found in Samuel van Quicchelberg's *Inscriptiones vel tituli theatri amplissimi* (1565) for forming and organizing a "paper museum" helped establish a new cultural role for prints. Quicchelberg recommended that sheets be stored in large albums, grouped by subject. To reduce problems in handling, he advised that acquisitions be limited to prints within certain dimensions.[1] He did not, however, suggest how to choose among prints, nor did he recommend printmakers. These omissions leave room to conclude that at this time authorship, condition, and impression quality were not primary concerns for collectors.

The Earliest Restoration Practices

Despite the generally low valuation placed upon prints in the sixteenth century and the interchangeability of prints of similar subjects, it is likely that some restoration was performed, prompted by either rarity or frugality. For those who assembled collections based on subject or for artists who used prints as models, the limited production of some subjects or compositions might prompt restoration when a valued impression was damaged.

Frugality was the more common impetus for a restoration treatment. If an artist or printseller could render a print salable by removing a spot or a crease, it would be natural to do so. Localized damage such as surface dirt, folds, creases, spots, and stains were the types of problems a restorer in a printing house would encounter. Artists had long used feathers or soft rags to brush away the dust of charcoal underdrawings, and these were also used to remove surface dirt from prints. Prints were routinely flattened in screw presses during production, and the use of these presses to remove accidental creases and folds were familiar to the printmaker. Stains created by the drippings of wax candles and oil lamps or by ink splatters presented a greater challenge,[2] but information on remedies was available.

Local stains from accidental spills and drippings were likely the targets of the earliest restoration treatments because—as opposed to the generally slower processes associated with overall discoloration—their effects are immediate and often far more distracting to the eye. The accident itself no doubt prompted the print's possessor to act.

The technology of stain removal from paper owes much to the treatment of textiles. Many methods for treating prints published in the nineteenth century were used in the sixteenth century for textiles. In late sixteenth-century books containing directions for spot removal from cloth, "paper" is sometimes slipped in among the clothstuffs. The late-Renaissance literature on stain removal from cloth is also the earliest literature on print restoration.

Early professional manuals for textiles, such as the *Plictho* (1548), contain instructions only for dyeing or processing.[3] Stain removal was largely a domestic art. Written instructions were commonly contained in "housebooks"—manuscript compilations of information relating to domestic management. Along with directions for spot removal, housebooks might include instructions for farming, purifying wine, cooking, defending the home, and stopping a bleeding wound. The tradition of the housebook can be traced from the Middle Ages through the late eighteenth century. Its utility was recognized by early book publishers, and the housebook evolved into the compendium of receipt, a miscellany of recipes and instructions that was very popular throughout Europe.

The first book on textile stain removal was published in 1532.[4] Numerous tracts containing practical information of this nature followed, including *A Profitable Booke* (1588), by Leonard Mascall. Typically the treatments outlined are for the removal of ink, grease, wax, and oil, using either poultices or solvents, depending on the source of the stain. These materials were also likely to be found as local stains on paper. One type of reagent bleach is discussed—salt and lemon juice.[5] By the end of the sixteenth century, and even more widely dispersed by the mid-seventeenth century in publications such as Giovanni Baptista della Porta's *Natural Magick* (1658) and Henry Peacham's *Compleat Gentleman* (1661), directions for stain removal, particularly for ink, were addressed specifically to paper.

Print Restoration and the Advent of Connoisseurship

Had collectors remained concerned solely with contemporaneous prints, satisfied with copies and impressions from reworked plates, and generally willing to replace damaged prints with new ones, the focus of print restoration might have stopped at local treatments. The rise of print connoisseurship brought a new purpose to print collecting. By establishing criteria for selecting prints and a program for grading impressions, print connoisseurship provided the stimulus for professional print restoration.

These concerns among print collectors, first evident in the late sixteenth century, centered on impression quality.[6] In the first decades of the seventeenth century a growing interest in physical condition also emerged.[7] These shifts in the concerns of print collectors coincide with an evolving recognition that prints could serve as illustrations for the history of art. The organizational schemes of some collections changed from subject matter to regional or national schools of art. In 1649 these developments were codified with the publication of Abraham Bosse's *Sentimens sur la distinction des diverses manières de peinture, dessein & gravure, & des originaux d'avec leurs copies*, a manual for connoisseurs that advised choosing prints from unreworked plates and that were well printed on paper that was "very white."[8]

After the mid-seventeenth century, developments in print collecting occurred more rapidly. Interest in proofs and states of prints grew.[9] Catalogues of collections were published. The use of the collector's mark became common, indicating that pride of ownership and a concern for provenance were increasing. However, prints continued to be thought of as vehicles for education, and they remained bound in volumes and kept in libraries where, as visual representations, they complemented texts. The focus of print collecting persisted in this tradition until Adam von Bartsch's *Le peintre-graveur* appeared in the first quarter of the nineteenth century. Bartsch's twenty-one volumes established a canon of desirable prints and promoted artists who designed and executed their works.[10]

Bosse's call for prints on paper that are "very white" is closely linked with the change in the purpose of print collecting that occurred in the early seventeenth century. The use of prints as representations delineating the history of art meant that collectors sought older, previously less-valued and less well cared for prints.[11] As a result of storage and display practices, some prints must have shown a degree of discoloration that the print connoisseur found unacceptable. The appearance of instructions for the overall bleaching of prints may be a response to the new function prints had in the collection.

A bookseller from the French village of Sedan provided the earliest known directions for the overall bleaching of prints. They are recorded in the famous manuscript, "Pictoria Sculptoria et Quæ Subalternarum Artium," compiled by the physicist and chemical investigator Théodore Turquet de Mayerne.[12] Little is known of the author, Jehan Anceaux, except that he was in London at the time that de Mayerne recorded the instructions, perhaps for the purpose of offering volumes to de Mayerne for his extensive library.

Dated 1631, these instructions for spot removal, resizing, and bleaching prints are the only ones directed specifically toward prints known to originate in the seventeenth century.[13] How widely employed these treatments were is impossible to determine, but they demonstrate that eighteen years before the publication of Bosse's guidelines the condition of the prints had become important to collectors. In the instructions, resizing paper with gelatin and alum was routine following any treatment involving aqueous immersion. The removal of ink was accomplished with *eau forte*, a generic term in the seventeenth century for acidified water but which may refer to the mordant used by etchers. Methods for handling wet paper were borrowed from papermaking and printmaking. To bathe a print, Anceaux proposed immersing the sheet in water in which a codfish had been boiled. Afterward, the print was to be removed and placed in water made alkaline by the addition of graveled ashes, an archaic term for potash, which would combine with the fish oil absorbed in the paper and form a soap. Sun bleaching, which entails dampening the print and exposing it to the rays of the sun, was recommended for further whitening.

Anceaux's hometown of Sedan was an important center in the seventeenth century for the production of textiles, and no doubt he was familiar with the practices of the linen bleaching fields that bordered the town.[14] His instructions further demonstrate the close relationship between textile processing and print restoration.

In the mid-eighteenth century, Paris was a center for the print and drawing trade. Collecting circles were dominated by Pierre-Jean Mariette, the fourth generation of a family of print publishers.[15] It was within this milieu that the book containing the first published directions for restoring prints, *Catalogue des estampes gravées d'après Rubens. Auquel on a joint l'oeurve de Jordaens, & celle de Visscher. Avec un secret pour blanchir les estampes & en ôter les taches d'huile* (fig. 1), was offered to the public in 1751. Robert Hecquet (1693–1775), the author, was active in numerous roles in the print trade. He was a printmaker (etching, among other works, a plate after Nicolas Poussin), a collector, and a dealer in prints. Hecquet's "secret" was sun bleaching, which by this time was better known as a restoration treatment than the title of the book suggests. Just as Jehan Anceaux related his instructions to de Mayerne, individuals with an interest in intaglio prints had passed along instructions for sun bleaching for more than a century. Perhaps more significant than the instructions themselves is Hecquet's effort to describe the nature and sources of the discoloration he attempted to remove. Two forms of discoloration were recognized: one, yellowish and situated along printed lines, was due to poorly processed ink; the other, reddish and more general, was thought to result from exposure to the air. These observations are among the earliest published attempts at understanding the causes of damage to art.

In 1767, Hecquet's instructions were republished with credit by Pierre Françoise Basan (1723–1797) in *Dictionnaire des graveurs*. Basan is best known today as a copyist of Rembrandt's prints and reprinter of his etching plates. In his own day, Basan was among the most important print dealers in Paris. Following Mariette's death in 1774, Basan handled the sale of Mariette's collection and inherited his position as the leading connoisseur of his time. Like Hecquet, Basan was probably versed in other print restoration techniques. Both may have sun bleached the older prints they sold and restored prints for others.

The Early Trade of Print Restoration

The trade of print restoration most likely began in a printing house where its knowledge would contribute to the number of salable prints. Once collectors became interested in older prints and concerned about their condition, restoration was encouraged. Stained and discolored prints could be purchased cheaply, restored, and then sold for considerable profit. Many of those engaged in collecting or printselling circles had access to the methods. An advertisement in the *London Evening Post* of 8–10 August 1710 underscores the connection between the printshop and the practice of print restoration and provides evidence of the broadening market for these services.

The Famous Battles of Alexander . . . curiously engraved by the most Famous Hands in Europe, in 13 Sheets of Paper, 3 Foot Deep, 2 Foot Broad each, fine French and Dutch Prints fit for Parlours, Stair-Cases and Closets, and likewise the Queen's Palaces, Noblemens Seats and Cathedrals curiously engraven; to be sold and fram'd very reasonable by Robert Hulton at the corner of Pall-Mall over against the Hay Market, where Gentlemen and Ladies may have their dirty yellow Prints very well cleaned and cheap.[16]

The market Robert Hulton attempted to reach through his advertisement was not comprised of print connoisseurs or dealers. His services, like the prints he sold, were for persons for whom prints served as wall decorations. By the beginning of the eighteenth century, print restoration had also become a domestic service. Note that Hulton did not refer to torn or spotted prints to be mended or cleaned. His services were confined to sun bleaching his clients' "dirty yellow Prints," discolored on exposure to light. It should also be noted that restoration was not the main thrust of Hulton's enterprise but a sideline.

The evidence regarding print restoration from the seventeenth to the mid-eighteenth century, while scant, suggests that the trade was a part-time endeavor. Dealers of older prints intended for the collector's market occasionally employed the secrets of print restoration as a means of improving their profits. Printsellers, engaged in vending contemporary impressions, also found knowledge of restoration techniques valuable in increasing their income. For printmakers and publishers the knowledge of restoration techniques was useful for repairing prints damaged by accidents in the shop or during transit.

A broadening interest in the collection and connoisseurship of art on paper in the later half of the eighteenth century changed the conditions that had maintained print restoration as a part-time enterprise. These changes were most notable in Paris, where several specialties emerged in the trades connected with art on paper. Mount making and framing developed to highly skilled levels. Drawn largely from the ranks of unsuccessful artists, one of the earliest professionals charged with the care of works of art on paper, the *dessinateur*, worked on large collections, creating mounts and performing restorations.[17] While no trade cards or bills of receipt have come to light to confirm that full-time print restorers were working for dealers in Paris or other European capitals at this time, conditions suggest their presence.

In an 1816 essay, George Cumberland described a product of the restorer's bench:

We have an instance before us in the ingenious manner in which a set of the Cupid and Psyche in the British Museum has been made to appear *perfect by uniting the written part of one impression with engraved part of another; and this so nicely done that nothing but the closest examination, with the help of a glass, enabled me to discover it, the tint being so exactly imitated, and the juncture so cleverly fitted, together with the imitations of the edge of the copper plate (by passing one through a press over them) that the deception was nearly completed.*[18]

It is possible the object described by Cumberland was an isolated case. Yet the manual skills and practical tricks necessary to unite two impressions require time to develop and once acquired would be used again. The appearance of such frauds at the beginning of the nineteenth century are proof that print restoration had evolved as a highly skilled craft, if lacking in professional ethics, and suggest again that restoration became an independent trade in the later eighteenth century.

It was also at this time that chlorine bleaches were discovered. The technology was transferred from textiles to print restoration by Joseph Chaptal in 1787.[19] Chlorine bleach became a favorite tool of a large class of print restorers, many of whom practiced restoration as a trade. It is tempting to conclude that the relationship between the rise of the trade of print restoration and the growing use of chlorine bleaches was complementary, that the presence of one encouraged the other. Yet the fine manual skills that characterize the best efforts of one style of print restoration contrast markedly in approach with the reliance of some restorers on chlorine bleaches. In the early nineteenth century, three distinct traditions, defined by intent and stylistic approach, become apparent within the general practice of print restoration.

The Traditions of Print Restoration

The three traditions of print restoration may be described by category of practitioner: the artist-restorer, the collector-restorer, and the domestic-restorer. Each tradition brought differing motivations, skills, and aesthetic concerns to the task of restoring a print. The artist-restorer sought to increase value. The collector-restorer's goal was preservation of the print and its attributes. Cleanliness was the primary concern of the domestic-restorer. These traditions form the main thrust of print restoration practice from the beginning of the nineteenth century until the 1930s, when the profession of art conservation and the introduction of the scientist-restorer began to influence print restoration practices.

The tradition of the artist-restorer naturally began with the restoration practices in the publishing and dealing houses. Many who practiced restoration as a trade in the nineteenth and early twentieth centuries were trained initially as printmakers or draftspersons. Their skills and knowledge of materials allowed them to perform many of the treatments that were the bane of collectors throughout the nineteenth and early twentieth centuries, including undetectable mends, deceptive inpainting, and the replacement of margins. At the same time collectors also often found the services of the artist-restorer useful. The efforts of the artist-restorer were nearly always market-driven, and many practices now considered unethical were considered acceptable if they increased the value of the print.

With the increased popularity of collecting in the early nineteenth century, the number of collectors who restored their own prints increased as well. There were several attractions, including one that is by now familiar: discolored impressions of desirable prints could be purchased cheaply and improved. However, collectors were often far from major cities where restoration services were available. Even if a restorer was available, the tradesperson could not always be trusted to proceed with the same restraint as the collector. The collector-restorer tradition is characterized by an informed aesthetic sense. The treatments performed and promoted were intended to replace or retain the subtle elements of a print, such as paper tone and surface. Long after the introduction of chlorine bleaches, sun bleaching remained the preferred manner of lightening a print's overall tone because the effect is limited and the result closer to the sheet's original color.[20]

The tradition of the domestic-restorer is defined by its heavy reliance and overuse of reagent bleaches and the tendency to equate discoloration in a print with dirt.[21] To the domestic-restorer, cleaning and bleaching a print were synonymous. For many the primary goal of a treatment was to return whiteness to the print or achieve a whiteness the print never had. Some in this tradition were collectors inattentive to the theories of connoisseurship. Many were tradespeople who received payment for their services. Others were simply well-intentioned cleaners or hobbyists. An early twentieth-century manual summarizes the motivations of this group: "One of the great charms of cleaning a print, quite aside from the pleasure it affords or the knowledge that it may considerably enhance its value, is the fact that it may be conducted at a very small outlay of materials."[22] Until the development of chlorine bleaches, this group was restrained by the limitations of sun bleaching. Reagent bleaching products provided a powerful tool whose small cost and dependable results had a particular appeal to the untheoried and manually unskilled. Beginning with the final decades of the eighteenth century, the wide availability of commercial restoration materials provided librarians, bookbinders, picture framers, housewives, and many others with the means to alter the appearance of a print to suit their taste.[23]

The stark white sought by the domestic-restorer is at odds with the values of the connoisseur, but one fact argues in this tradition's defense. The history of print collecting and scholarship is marked by a continuous reexamination of the body of available prints. Through this process a reassessment of value is given certain works, changing their importance to collectors.[24] When the prices of one class of prints rise or impressions become scarce, collectors frequently seek out other, less-appreciated areas of prints to satisfy their acquisitive nature. A print initially chosen for its low cost and pleasing image and eventually bleached to stark whiteness while serving as a wall decoration might later be reappraised as a valuable impression and the restorer condemned for ruining a fine print.

Print Restoration Literature

As the popularity of print collecting increased, the numbers of fine impressions in excellent condition, always scarce, decreased. Many prints passed permanently from private hands into the expanding collections of institutions. Higher prices and greater demand brought more prints of lesser quality to the market; many of them required restoration treatment to make them desirable. Contributing to the growth of print restoration and its literature was the shrinking size of the average print collection. In contrast to Michel de Marolles, who in the seventeenth century amassed two collections each containing more than 100,000 sheets, private collections of prints would rarely again number in the tens of thousands.[25] With fewer prints to care for, the owners of smaller collections focused more attention on mounting and storage, and the print in less than optimum condition was more likely to receive a restoration treatment.

The first major tract on print restoration, *Vollständige Anleitung zur Erhaltung, Reinigung und Wiederherstellung von Gemälden, Aquarellen, Kupferstichen, etc.* was published in 1812 by Friederich Lucanus, an apothecary from Halberstadt. Little is known of Lucanus' life, and it is not clear whether he was a practicing restorer or an informed amateur engaged in the arts. He was the first person, however, with a background in science to write at length on matters of print restoration and the first to describe dammar varnish in scientific literature.[26] His slim volume on the restoration of paintings and artwork on paper went through numerous editions over more than 150 years (see appendix). The influence of Lucanus' work was widespread and is reflected in its incorporation into the first Polish manuscript on print restoration, written in 1840.[27]

In the active collecting milieu of Paris, Alfred Bonnardot's (1808–1884) *Essai sur la restauration des anciennes estampes et des livres rares* was first issued in a limited edition in 1846. In 1858, a revised edition was published as *Essai sur l'art de restaurer les estampes et les livres*, followed in 1859 by a German translation. This was the most important restoration text published in the nineteenth century. Bonnardot credited picture framers as the stimulus to his undertaking, whom he likened, through their heavy reliance on reagent bleaches when cleaning prints, to those who would whitewash ancient cathedrals.

Bonnardot's book was the result of many years of study and research. In 1838, he commenced to collect drawings, prints, and plans relative to the ancient monuments and historic events of his beloved Paris. Even earlier, beginning in 1829, Bonnardot had begun to attend the scientific lectures of Joseph L. Gay-Lussac and Louis J. Thénard. With his collection and a grasp of scientific principles, Bonnardot carried out the practical experiments that formed the basis of his book. Following its publication, a would-be restorer had for the first time not only instructions for restoration methods but recipes for materials and information about equipping and organizing a workshop.

With this information alone, Bonnardot's contribution is substantial. However, he extended his discussions beyond the physical necessities of print restoration to establish criteria for when artwork should or should not be treated. He recognized that some forms of physical damage represent documentary evidence. Stated in its simplest form, he recommended that damage resulting from time or negligence should be repaired, but damage produced by use, such as the worn pages of a fifteenth-century prayer book or marginal inscriptions of a previous owner, should remain; to do otherwise would rob the article of its "poetry." It is the development of restoration theory that sets Bonnardot's book apart from all other books on print restoration produced during the period covered in this study.

Significantly, when Bonnardot began revisions on his book, he returned to his experiments and thus was able to observe the effects that more than ten years of aging had on the samples. With this concern for the effects of treatment over time, his background in science, and recognition of the role of connoisseurship and treatment theory, Bonnardot developed many of the principles that inform modern fine art conservation. He deserves recognition as the first modern paper conservator.

The *Essai* of 1846 was not Bonnardot's first published work. *Perruque et noblesse: Fatalité en trois parties* was published in 1837 and printed in a limited edition of five hundred, of which at least half were burned in disgust by the author.[28] Perhaps not coincidentally, the author's next undertaking was the *Essai*. The years between the two editions of the *Essai* appear to be the author's most productive period. The titles of his other works suggest the range of his interests. Some, such as *Dissertations archéologiques sur les anciennes enceintes de Paris* (1852) and *Historie artistique et archéologique de la gravure en France* (1849), were based on his collecting concerns. Bonnardot combined this interest with an active imagination to write two novels set within collecting circles, *Le mirourer du bibliophile parisien* (1848) and *Le povrtraict de l'iconophile parisien painct au vif* (1852). Other publications provide a glimpse of Bonnardot beyond the world of art; in 1855, he authored *Des télescopes*, and in 1856, he wrote a small article about lapdogs.[29]

From Bonnardot's writings, an image of the man emerges. One imagines a smallish, energetic man, described as eccentric and possessing a misanthropic sense of humor;[30] a lover of old books and prints whose time away from these devotions is spent gazing into the night sky or settled into a stuffed chair with a miniature spaniel on his lap. However, the most enduring image, and his lasting legacy, is Bonnardot the collector-restorer who, inspired by a love of the printed word and image on paper, trained himself in print restoration and wrote a book that was a tremendous leap beyond anything preceding it.

Other authors made valuable contributions to the literature of print restoration, and a look at their backgrounds provides insight into print restorers as individuals. Joseph Frederick August Schall (1785–1867) fits neatly into the profile of artist-restorer. Schall trained as an artist and was best known for painted miniatures and drawings in chalk and in ink. He was also a lithographer and an engraver specializing in reproductive work. He credited the chance purchase of some prints at auction in 1812 as the stimulus for learning restoration, and he relied on the first edition of Lucanus' book, published the same year, for his initial guidance. Schall offered his restoration services for payment, restoring thousands of prints in his lifetime, and he taught some of his techniques to other artists. How he divided his time between artistic pursuits and print restoration is unclear, but Schall must certainly have found his training as a miniaturist and reproductive engraver compatible with the precision required for print restoration.

Schall's manual, *Ausführliche Anleitung zur Restauration vergelbter, fleckiger, und beschädigter Kupferstiche u.s.w.* was published in 1863. While less thorough than Bonnardot's book in some respects, Schall's manual covered subjects omitted by Bonnardot, such as inpainting portions of the image that are lost or worn and the aesthetics of margin trimming. Perhaps to make the manual more attractive to print collectors, Schall included instructions for making etchings and aquatints and a catalogue of the prints he considered important. Most of the prints cited are reproductive works.

By the mid-nineteenth century five major tracts on print restoration had been published in northern Europe. The works by Hecquet and Bonnardot were available in French. Those by Lucanus, Bonnardot, and Schall were available in German.[31] No English-language manual for the restoration of prints paralleled the scope of the continental examples until 1911, when *Print Restoration and Picture Cleaning* by Maurice James Gunn was published. In England this void was filled by print collector's manuals, which often contained restoration instructions in addition to catalogues of printmaker's works, information on the examination and selection of prints, and directions for their mounting, display, and storage. By offering a different perspective, these manuals are valuable in serving to complete our picture of the development of print restoration and the goals of restoration treatments.

A surprising number of print collector's manuals published in the nineteenth century contain instructions for restoring prints. Among them are J.C.L. Musseau, *Manuel des amateurs d'estampes* (1821); George Cumberland, *An Essay on the Utility of Collecting the Best Works of the Ancient Engravers of the Italian School* (1827); Joseph Maberly (first published anonymously), *The Print Collector* (1844), published in a later edition (1880) with additions by Robert Hoe; William Willshire, *An Introduction to the Study and Collection of Ancient Prints* (1877); Andrew Tuer, *Bartolozzi and His Works* (1881); George Slater, *Engravings and Their Value* (1891); and Cyril Davenport, *Mezzotints* (1904). Some of these books, such as Tuer's *Bartolozzi and His Works* and Davenport's *Mezzotints*, display a promotional tack, providing restoration techniques because the initiate to the newly described area of print collecting would discover many prints in less than optimum condition. As a group, the collector's manual offered alternatives to those who might wish to bypass dealers and agents and purchase prints at auction. The inclusion of restoration techniques in collector's manuals suggests there were still relatively few independent print restorers whose services were available to collectors, as for the most part the trade restorer worked for dealers.

In general, collector's manuals are conservative in their approach to print restoration. They stress the connoisseur's interest in retaining valued elements of the print, warnings of the risks of treatment, and concern for the long-term preservation of the collection. Sun bleaching continued to be mentioned as a restoration technique in this genre of literature, but reagent bleaches were strongly condemned, particularly in the first half of the nineteenth century. Later, the moderate use of reagent bleaches was proposed, always to be followed with cleansing rinses to wash out residual chemicals that would shorten the life of the print.

Another form of literature containing instructions for print restoration was more popular than either the restoration manual or the print collector's manual. The compendium of receipt was designed to engage the domestic, the frugal, and the self-reliant. Tucked among recipes for shoe polish and advice for curing a drunk are directions for print cleaning and occasionally other treatments such as flattening, washing, mounting, and framing. The compendia were responsible for widely dispersing instructions for cleaning prints by immersion in a reagent bleach. Directions rarely include warnings against overbleaching or residual chemicals. The domestic-restorer found compendia particularly useful.[32]

A specialized form of the compendium of receipt is the printer's or artist's handbook. These sources—collections of recipes for the arts and gleanings from other trades and sciences—differ from the artist's manuals that stressed artistic technique and theory.[33] The inclusion of instructions for bleaching prints in *The American Artists' Manual* (1814) by James Cutbush and the very direct and practical information on bleaching, spot removal, and the flattening of creases found in Charles Jacobi's *Printers' Handbook* (1891) give evidence that traditions of print restoration were alive in printing houses and artists' studios throughout the nineteenth century.

The first English-language manual for print restoration testifies to a growing hobbyist's interest in the field. *Print Restoration and Picture Cleaning* by Maurice James Gunn was published in 1911. Gunn stressed the pleasure and profitability of print restoration, promoting it as a domestic enterprise that women should learn:

> *Nor is the work of a kind suited only to one sex—men. There is not the slightest reason why women should not prove quite as capable as men. Indeed I would go further and say that much of the work entailed is, if anything, better suited to the more delicate handling of women than men, and there is not the slightest reason why once the ABC of the subject has been thoroughly learned, women should not, equally with men, enjoy the reputation of being capable restorers, and by such means add to their incomes.*[34]

Gunn offered the reader some of the era's most heavy-handed print restoration practices combined with some of the most traditional. He emphasized expedience over manual skills and advocated squaring losses (trimming the edges of losses) to make clean fills and the use of the screw press for flattening. While warning against overcleaning, he relied completely on reagent bleaches. He also discussed hanging prints over a line to dry, a technique little mentioned in the literature of print restoration since the days of Jehan Anceaux. Chapters on fakes, print values, and prints to collect were addressed to collectors, but they are only supporting material for the primary focus on print restoration.

Print Restoration and Picture Cleaning presents the most severe approach in the print restoration literature to 1930. The favored bleaching reagent was chloride of lime; no mention was made of sun bleaching. To remove obstinate stains, Gunn suggested overbleaching followed by toning and resizing. His preferred method of mounting was inlaying.[35]

Print Restoration and Picture Cleaning

was published in a second edition in 1922. Similar instructions are found about this time in publications in France, Germany, and England. Gunn's book may have been so influential that it solidified the literature of print restoration, but more likely it simply reflected the current approaches.

Methods of Producing and Preserving Prints by C.W.F. Goss followed many of Maurice Gunn's methods. Published in 1915, the manual originated as a paper read to a library association. It is notable for its support of grangerizing.[36] Goss wrote many other books, none concerning prints or print collecting.

Restoration instructions appeared in a 1924 collector's manual, *Der Kupferstichsammler*, by Heinrich Leporini of the Albertina Museum, Vienna. In keeping with the tradition of the collector's manual, sun bleaching was mentioned, but without the enthusiasm of its earlier promoters. The author was clearly versed in the literature of print restoration. While his three-entry bibliography of books on restoration included Bonnardot, Schall, and Lucanus,[37] the restoration methods he outlined were taken from Maberly, Davenport, and Tuer. Like other informed collectors, Leporini stressed the importance of retaining the paper's original tone and surface. Yet, reflecting current approaches, he also explained how to split a sheet of paper and remargin a print.

The last important collector's manual in this survey is *Therapeutica graphica* (1930) by Frédéric-Charles Lonchamp. This manual covers the collection and restoration of books, pastels, watercolors, drawings, and prints. The instructions for restoring prints are characteristic of many restoration procedures of the era. The only bleaching agent described is calcium chloride. The author recommended and offered instructions for remargining; he also gave directions for replacing the embossment of the plate mark and for transferring missing portions of images from other impressions of the same print. His mention of Japanese paper for mending is an early reference to this now-common practice. His designs for storage shelves and boxes for prints reflect a concern for preventing further damage to a collection.

In general, published instructions for print restoration were written for the benefit of the amateur. Restorers involved in the trade did not write for their peers; the audience was too small and there was no advantage in divulging one's secrets to the competition. For this reason there are no published instructions from the nineteenth century for the fraudulent styles of restoration common in the period.

Collectors usually provided the most balanced discussion of treatment options and risks. It is often difficult to determine the source of the collector-restorer's information. Much seems to have been obtained through individual trial and error and through correspondence with other collectors. In the early twentieth century, instructions for print restoration appear less frequently in collector's manuals. Those that were published tend to be more intrusive and lack the cautious tone of instructions published earlier. Increased access to the services of print restorers may have played a part in the disappearance of all but the simplest instructions for mounting and restoration. Slater's *Engravings and Their Value* documents this trend. When first published in 1891, this book contained some of the most severe restoration instructions found in print-collecting literature. In the 1912 edition, the instructions were dropped.[38]

The Print Restoration Trade

Like most art restoration trades, print restoration, although a skilled craft, was for most of its history undocumented. There is little information concerning training. In the seventeenth and eighteenth centuries restoration was probably included in training in the print trades. An assistant in the publishing or dealing house who displayed the proper degree of care in the performance of routine duties may have been chosen to learn the more exacting craft of restoration. Andrew Tuer's reference in 1881 to the restorer Grisbrook as descended from the restoration shop of William Baldwin, who was in turn a former employee of the printseller Holloway, is the best-documented example.[39] Many who gravitated into print restoration, most notably Alfred Bonnardot and Joseph Schall, were self-trained.

Schools for restorers were a late development. They are less a part of the print restoration tradition than evidence of the developing profession of art conservation. In 1924, Heinrich Leporini referred to the school established at the Berlin Printroom by "the restorer Houser."[40] The establishment of restoration workshops in institutions opened many new opportunities for training restorers. Institutions had the resources for operating schools and student assistants to help maintain large collections.

The comments of collectors offer a perspective on the development of the print restoration trade. In 1816, George Cumberland had warned his fellow collectors of the mischief they might encounter in prints that had passed over the restorer's bench, and he accused restorers of innumerable duplicitous acts. Through much of the nineteenth century, Cumberland's attitude toward print restorers was perpetuated in collector's manuals. By the end of the century, however, while still warning collectors of the potential of fraud when purchasing prints, these manuals also stressed the importance of knowing a good print restorer and often recommended restorers by name. Tuer recommended "Mr. Grisbrook of Panton Street, Haymarket, who has been in the business over thirty years." Grisbrook is, said Tuer, "perhaps the best living restorer and inlayer of prints, and when anything very special is required, his are the services generally sought." Tuer described Grisbrook's professional lineage, the articles in his shop, and some of his practices. Tuer still warned the reader away from the unskilled or insensitive "so-called restorer" and the products of the unscrupulous.[41] It is evident that by the early 1880s, collectors had begun to patronize restorers.

The collecting climate of the nineteenth century was conducive to the establishment of independent print restorers. Throughout this century and into the next, new areas for acquisition and scholarship opened up, and each brought previously unappreciated prints to the attention of informed circles. Ambitious restorers saw the opportunity for occupational independence. By positioning themselves between the collector and the dealer, they could provide services to both ends of the market. Many were also still involved in the production or sale of prints.

The increased familiarity of the collecting public with print restoration moderated attitudes toward the print restorer's activities. Some secrets of the trade became less mysterious; eventually they would appear as recommended techniques in collector-authored restoration manuals.[42] Some collector's manuals stressed the legitimacy of the restorer's craft. Alfred Whitman of the department of prints and drawings of the British Museum made this point with evident reservations: "The collector will need to know an experienced restorer to whom he can entrust his treasures to be cleaned and repaired, and *care must be taken that the work done to the prints is not carried too far.*"[43]

Little is known about fees charged for restoration services, but two examples from the early twentieth century illustrate their range. In Chicago, in 1915, a print could be bleached and matted in the shop of one of the city's most prominent art dealers for one dollar. In the same gallery, a mat by itself cost seventy cents.[44] The difference of thirty cents attests to the small cost and cursory nature of reagent bleaching treatments in this period.

The example demonstrating the other extreme requires more explanation. In 1931, Lessing J. Rosenwald's impression on oriental paper of the *Hundred Guilder Print* by Rembrandt van Rijn was on an exhibition tour in museums in the eastern United States. While in transit between venues, the print received a scratch on the image approximately three-quarters of an inch long from a "cinder" that had managed to find its way beneath the glass. In his search for a restorer, Rosenwald sought the advice of the country's leading expert on drawings and prints, Paul Sachs of Harvard University's Fogg Art Museum. Sachs recommended Carl Schweidler of Berlin who, with his brother Max, operated the most famous (and at times infamous) print restoration shop in Europe. It is said that the brothers began as printmakers. Many well-crafted and deceptive restorations of prints in museum collections are today credited to this workshop.

After corresponding with Schweidler, Rosenwald had the print shipped to Berlin to have the short, superficial scratch inpainted. For this work, requiring at most fifteen to twenty minutes to complete, Carl Schweidler charged Lessing Rosenwald four hundred dollars.[45]

This story illustrates several points. First,

it shows that in 1931 Paul Sachs knew of no one in the United States to whom he would entrust the rather routine yet delicate task of inpainting a thin scratch of less than one inch. Several more years would pass until the first conservation laboratories were established in museums in the United States.[46] It is possible that Schweidler based his fee, which in depression-era dollars was very high, in part on the client's ability to pay. However, it is rumored that Schweidler charged his clients for the restoration of value. In a letter Schweidler assured Rosenwald that "the sheet will have the same value as heretofore before it was scratched." No mention was made of an improvement in the impression's appearance.[47]

The Emergence of Paper Conservation

A new tradition arose in the first decades of the twentieth century that was to change the course of professional print restoration and develop eventually into the profession of art conservation. Here restoration laboratories in museums and scientific research into the materials of art and restoration played an important role. In the late nineteenth century and the early decades of this century, government agencies acknowledged the importance of preserving works of art and documents and began to fund research in these areas.[48]

The results of scientific research into the preservation of works of art are seen in *The Cleaning and Restoration of Museum Exhibits*, published by the British government's Department of Scientific and Industrial Research in 1921. Based on research performed by Alexander Scott, this bulletin placed an emphasis upon scientific principles not seen in restoration literature since Bonnardot. It recommended what would become a standard practice in art conservation: the use of before-and-after photographs to document the effects of treatment.

Expanded editions of *The Cleaning and Restoration of Museum Exhibits* were published in 1923 and 1926. The bulletin became an influential treatise on the restoration of prints, and its recommendations appeared in other books soon after it was first published. In one of the best-written and earliest printmaking manuals to discuss restoration techniques, *The Art of Etching* (1925), the British printmaker E. S. Lumsden drew extensively upon this British Museum publication. Jan Poortenaar also found Scott's research useful and drew from it in his book on etching, *Etkunst*, published in 1930.

The effect of the application of scientific principles to the treatment of art can be seen in Thomas R. Beaufort's popular manual, *Pictures and How to Clean Them*, published in 1926. Beaufort was a professional restorer and an adviser to the *Art Trade Journal*. The manual concerns the restoration of nearly all forms of pictorial art, from paintings to pastels. While Beaufort's approach in the treatment of prints differs little from earlier restoration manuals, his chapters on cleaning varnished and Japanese prints cover relatively new topics. A marked change from previous popular manuals is evident in his concern for the quality and purity of the materials used during treatment. In addition to providing a working discussion of chemical principles, Beaufort warned that the surface qualities of some papers may change due to treatment. A chapter on photographing art is included but without mention of its use for documenting treatments.

Two texts published in the late 1930s deserve mention for their striking differences and similarities. *The Conservation of Prints, Drawings, and Manuscripts* represents a continuation of research at the British Museum. The author, H. J. Plenderleith, was both assistant keeper in the British Museum Laboratory and professor of chemistry at the Royal Academy of Art. His book, published in 1937, is the first on the subject of the physical care of art on paper to include "conservation" in its title. In keeping with this philosophical change, there is a lengthy discussion of the causes and prevention of damage and proper methods of storage and display.

Max Schweidler's *Instandsetzung von Kupferstichen, Zeichnungen, Büchern usw*, published in Berlin in 1938, is the most complete publication coming from the tradition of print restoration. It includes instructions for splitting prints and false watermarking[49]—practices that compromise the physical integrity of a print and are used to create frauds. However, Schweidler promoted sun bleaching, while the chemist Plenderleith encouraged the use of reagent bleach.

While Plenderleith and Schweidler differ in approach, both, for the first time in restoration texts, discuss how paper is made. Both discuss printmaking processes and drawing media. Both identify differences in paper and present treatments designed to accommodate the variations. In both traditions—the old-school print restorer and the new, science-based profession of art conservation—there is a growing recognition of the subtleties involved in the treatment of works of art on paper and an understanding that success begins with a thorough understanding of the materials and methods used in their creation.

Conclusion

The rise of the profession of art conservation in the early twentieth century, with its regard for the long-term physical stability of art, insistence on scientific principles, and emphasis on ethical standards signaled a redirection in the tradition of art restoration. Yet, for the previous hundred years these same concerns had been embraced by some of those who restored prints. Today the traditions and practices of print restoration still remain in many ways. They affect the modern profession of paper conservation in its recognition of the role of connoisseurship in the treatment of prints and in its admonitions against conservators' dealing in art.

The practice of print restoration left a legacy that has been often disparaged and frequently ignored. Yet, in workshops and in the backrooms of frame shops in Europe and the United States, many of the practices described above continue. It is hoped this survey can clarify some of the motivations, issues, and practices that contributed to the development of print restoration, its literature, and its traditions.

NOTES

The information contained in this article was compiled during my tenure as Mellon Fellow in paper conservation at the National Gallery of Art, Washington, from September 1988 to August 1990. Many helped to make this article possible, and it is a pleasure to be able to recognize their contribution here. I wish to thank Tawney Becker, whose assiduous translations of many of the restoration instructions on which this research is based contributed immeasurably to the content of this paper. Among the many people who offered suggestions or provided encouragement at the National Gallery of Art, I thank, in particular, Shelley Fletcher, head of the paper conservation department; Pia DeSantis Pell, former senior paper conservator; Ross Merrill, chief of the conservation division; and Michael Skalka, conservation administrator. I thank the Mellon Foundation for providing the funding for my fellowship. I also appreciate the assistance of the following individuals who provided sources, fielded questions, and encouraged this pursuit: Marjorie B. Cohn, E. René de la Rie, Christopher Foster, Christa Gaehde, Robien van Gulik, Jan Johnson, Theo Laurentius, Liz Lunning, Jane McAusland, Katerina Mayer-Haunton, Roy Perkinson, Peter Poldervaart, William Robinson, Gisela van Rossum, Philip Stevens, Carel van Tuyll van Serooskerken, and Bas van Veltzen.

1. Elizabeth M. Hajós, "The Concept of an Engravings Collection in the Year 1565: Quicchelberg's 'Inscriptiones vel tituli theatri amplissimi,'" *Art Bulletin* 40 (1958), 151, 152.

2. A nineteenth-century reference to these stains is from the print publisher Rudolph Ackerman who, in a book on gas lighting published by his firm, mentioned the large number of prints lost in his printshop due to drips of oil and tallow from lighting devices. See Frederick Accum, *A Practical Treatise on Gaslight* (London, 1815), 72.

3. Gioanventura Rosetti, *Plictho de larte de tentori che insegna tenger pani telle banbasi et sede si per larthe magiore come per la comune*, 1548, trans. Sidney M. Edelstein and Hector C. Borgetty (Cambridge, Mass., 1969).

4. Moshe Ron, "The Sidney M. Edelstein Collection of the History of Dyeing, Bleaching, and Dry-Cleaning of Textiles," *Textile History* 12 (1981), 120.

5. The combination of salt with lemon juice would generate a mild solution of hydrochloric acid.

6. Michael Bury, "The Taste for Prints in Italy to c. 1600," *Print Quarterly* 2 (March 1985), 17.

7. William W. Robinson, "'This Passion for Prints': Collecting and Connoisseurship in Northern Europe during the Seventeenth Century," in Clifford Ackley, *Printmaking in the Age of Rembrandt* [exh. cat., Museum of Fine Arts] (Boston, 1981), xlii.

8. Robinson 1981, xli.

9. Robinson 1981, xl–xli.

10. Hajós 1958, 156. See also Walter Koschatzky,

"Adam von Bartsch: An Introduction to His Life and Work," in *The Illustrated Bartsch* (New York, 1978), 1: xvi.

11. Possible exceptions are prints by Albrecht Dürer and Lucas van Leyden, which had been eagerly sought throughout Europe in the sixteenth century. More than for most other printmakers, the delicacy of their burin work is diminished by darkened paper tone. Prints by these artists may have been the subject of bleaching treatments at a date prior to that proposed here.

12. Théodore Turquet de Mayerne, comp., "Pictoria Sculptoria et Quæ Subalternarum Artium," 1620, British Museum, London, MS Sloane 2052. The manuscript is published in Ernst Berger, *Quellen für Maltechnik* (Munich, 1901), original text and German translation, and as *Le manuscrit de Turquet de Mayerne* (Lyon, 1974), original text and French translation.

13. For a more complete discussion of these instructions, see Mark Stevenson, "A Seventeenth-Century Manual for the Restoration of Prints," *Print Quarterly* 7, no. 4 (December 1990), 420–424.

14. De Mayerne, 1620, fol. 61. The author's instructions are to "lay [the print] out on the grass or straw . . . as one does with linen."

15. Maxime Préaud, Pierre Casselle, Marianne Grivel, and Corinne Le Bitouzé, *Dictionnaire des éditeurs d'estampes à Paris sous l'Ancien Régime* (1987), 155–156.

16. Quoted in William Bragg Ewald Jr., *Rogues, Royalty, and Reporters: The Age of Queen Anne through Its Newspapers* (Westport, Conn. 1978), 179. John Hulton, perhaps a son, had a printshop at or very near the same location, "on Pall-Mall where it crosses Hay Market," in the 1730s. See Herbert M. Atherton, *Political Prints in the Age of Hogarth* (Oxford, 1974), 23.

17. Joseph Meder and Walter Ames, *The Mastery of Drawing* (New York, 1978), 479, 491.

18. George Cumberland, *An Essay on the Utility of Collecting the Best Works of the Ancient Engravers of the Italian School* (London, 1827), 23, Cumberland's italics. Although published in 1827, the essay was written in 1816. Unfortunately Cumberland failed to mention the artist.

19. Joseph Chaptal, "Observations sur l'acide muriatique oxygenee (chlore liquide) applique au blanchiment des estampes," *Annales de Chemie* 1 (1787), 69.

20. References to sun bleaching appear throughout the correspondence and writing of eighteenth- and nineteenth-century print collectors. One example is especially interesting for its relatively early date and the apparent precocity of its practitioner. In his autobiographical writings, Johann Wolfgang von Goethe refers to his youthful experiences as a print restorer. With the help of his sister, he laid prints belonging to his father on boards, wetted them, and propped them in the roof gutters so they could bleach in the sun. Afterward, the prints were taken to a bookbinder, who mounted the prints, "straightened the edges," and bound them into a volume. Johann Wolfgang von Goethe, *Werke*, Hamburger Ausgabe, 12 vols. (Munich, 1981–1986), vol. 9, bk. 4, 122.

21. I use the term "reagent bleach" to describe bleaching methods that rely on a chemical additive to induce the bleaching process as distinguished from sun bleaching, which, while a chemical process, employs the energy of light.

22. Maurice James Gunn, *Print Restoration and Picture Cleaning* (London, 1911), 3.

23. In the seventeenth century, poulticing materials for the removal of grease were available from apothecaries. By the early nineteenth century, solvent preparations for removing oil and grease were sold by color makers, rubber erasers could be bought from stationers, and chlorine bleaches were generally available.

24. Examples of this process are evident in the increased recognition in the first half of the nineteenth century of the importance of early Italian engravings and a similar awakening of interest in early Germanic engravings, which occurred slightly later. See Evelyn Erlich and John H. Neff, *The Gott Impression of Pollaiuollo's Battle of the Nudes* [exh. cat., National Gallery of Art] (Washington, 1973), 7; Jan Piet Filedt Kok, ed., *The Master of the Amsterdam Cabinet, or the Housebook Master, ca. 1470–1500* (Princeton, 1985), 42–43.

25. Louis B. Metcalf, "A Prince of Print Collectors: Michel de Marolles, Abbé de Villelion," *Print Collector's Quarterly* 2 (October 1912), 316–340.

26. Friederich Lucanus, "Dammar Resin: Concerning the Properties, Chemical Behavior, and Technical Application," *Schweigger's Journal* 55 (1829), 60–61. See also Robert Feller, "First Description of Dammar Varnish Translated," *Bulletin of the American Group-IIC*, 7, no. 1 (1966), 8, 20; Feller, "Quarterly Report of National Gallery of Art Research Project No. 4345–17," from 1 November 1966 to 31 January 1967, 12.

27. Władysław Slesiński, "O niezanym wkładzie Karola Soczyńskiego do nauki o konserwacji malowideł i grafiki," *Ochrona Zabytkow* 26 (March 1973), 176–180.

28. Alfred Bonnardot, *The Mirror of the Parisian Bibliophile*, foreword and trans. T. W. Koch (Chicago, 1931), xx. The source of Bonnardot's disgust is not clear.

29. Georges Vicaire, *Manuel de l'amateur de livres du XIX siècle, 1801–1893*, 7 vols. (Paris, 1894–1920), 1:850–855.

30. Bonnardot 1931, xxvi.

31. A fourth German publication on print restoration may have been available at this time. The fifteen-page pamphlet, *Deutliche und vollkommen bewährte Anweisung, aus Büchern, Stahlstichen, Kupferstichen, Lithographieen, und Holzschnitten, Flecken aller Art . . . zu vertilgen*, by E. F. Balinger, is undated.

32. A favorite volume of mine is titled *How to Make Home Happy* (1852) by William A. Jones, conjuring

images of domestic serenity achieved through stark white prints.

33. With few exceptions, such as Richard Tothill's *Arte of Limning* (1573), which contains one of the earliest published instructions for removing grease from paper, artist's manuals rarely contain information relating to the restoration of works of art.

34. Gunn 1911, 3–4.

35. Inlaying is similar to false margining. A window slightly smaller than the print is cut into the mounting sheet. Generally there is an overlap along the edges of the print and the mounting sheet and no intent to deceive.

36. Grangerizing is adding extra illustrations to a book by cutting other books or periodicals apart for their photographs or engravings. This practice originated with the publication in 1775 of the Reverend James Granger's *Biographical History of England* and produced a rage for providing supplementary illustrations of persons and places being discussed. Many prints were destroyed in the process. Some grangerized books display considerable skill in their construction, suggesting professional restorers did the work.

37. These restoration texts were either published in Germany or available in German translation.

38. George Herbert Slater's *Engravings and Their Value* (London, 1891), included instructions for toning prints after reagent bleaching by smoking them over a greenstick fire. "Punching" was recommended for removing foxing stains—small circular reddish brown areas of discoloration. In this method the restorer strikes the damaged area with a metal punch and replaces the ensuing loss with paper pulp. It is possible that the disappearance of these and other restoration instructions from the 1912 edition of *Engravings and Their Value* was due in part to the publication of Gunn's *Print Restoration and Picture Cleaning* the previous year by the same publisher, L. Upcott Gill.

39. Andrew Tuer, *Bartolozzi and His Works* (London, 1881), 92.

40. Heinrich Leporini, *Der Kupferstichsammler* (Berlin, 1924), 130.

41. Tuer 1881, 92.

42. Directions for false margining, replacing plate marks, and splitting prints appear in the collector's manuals by Lonchamp (1930) and Leporini (1924).

43. Alfred Whitman, *The Print-Collector's Handbook* (London, 1901), 104, Whitman's italics.

44. Art Institute of Chicago, print collection archives.

45. This information is based on correspondence in the National Gallery of Art, Washington, archives. The print, now in the collection of the National Gallery, has been examined in the paper conservation laboratory. An inpainted scratch, matching the length and possibly resulting from the source described by Rosenwald, is located just above and to the right of the head of Christ.

46. For information regarding the growth of art conservation in the United States, see Elisabeth C. G. Packard, "Conservation Goes Public: George Leslie Stout and the Art Technical Sections," part 1, *American Institute for Conservation Preprints*, 7th Annual Meeting (Washington, 1979), 86–93; Packard, "George Stout and the Art Technical Sections of the AAM Annual Meeting in the 1930's and 1940's," part 2, *American Institute for Conservation Preprints*, 8th Annual Meeting (Washington, 1980), 72–83.

47. Carl Schweidler to Lessing J. Rosenwald, 10 June 1931, National Gallery of Art, archives.

48. Britain and Austria entered into this research at an early date. By the mid-1920s the United States government also funded investigations of the preservation of paper and documents.

49. Paper splitting is usually done in the process of false margining a print. The most common method is to paste fine cotton duck or union cloth onto each side of the print and allow the whole to dry under pressure. When dry, a corner of the print is nicked, and with a quick, even pull, the paper is divided in two sheets, one with the image and one without. False watermarking is the practice of placing a watermark into the paper of a print, generally by cutting partway through the paper. The purpose is most often to make a later impression appear to be an earlier one.

APPENDIX

Sources of Instructions for Print Restoration from the Late Renaissance to the 1930s

Andés, Louis Edgar. *The Treatment of Paper for Special Purposes*. London, 1907.

The Artist and Tradesman's Guide. Utica, N.Y., 1827.

The Artist's Companion and Manufacturer's Guide. Boston, 1814.

Balinger, E. F. *Deutliche und vollkommen bewährte Anweisung, aus Büchern, Stahlstichen, Kupferstichen, Lithographieen und Holzschnitten, Flecken aller Art . . . zu vertilgen*. Stuttgart [mid-19th century?].

Basan, Pierre Françoise. *Dictionnaire des graveurs anciens et modernes depuis l'origine de la gravure avec une notice des principales estampes qu'ils ont gravées. Suivi des catalogues des œuvres de Jacques Jordans, & de Corneille Visscher*. 3 vols. Paris, 1767.

Beal, Mary. *A Study of Richard Symonds*. New York and London, 1984.

Beaufort, Thomas R. *Pictures and How to Clean Them*. London, 1926.

Beman, David. *The Mysteries of Trade*. Boston, 1825.

Bonnardot, Alfred. *Essai sur la restauration des anciennes estampes et des livres rares, ou, Traité sur les meilleurs procédés à suivre pour réparer, détacher, décolorier et conserver les gravures, dessins et livres*. Paris, 1846. *Essai sur l'art de restaurer les estampes et les livres, ou, Traité sur les meilleurs procédé pour blancher, détacher, décolorier, réparer et conserver les estamples, livres et dessins*. 2d rev. and enl. ed., Paris, 1858. German trans. Quedlinberg, 1859. English trans. of unpublished typescript by William C. Hollands. Ann Arbor, Mich. [1932?].

British Government, Department of Scientific and Industrial Research. *The Cleaning and Restoration of Museum Exhibits*. Bulletin 5. London, 1921, 1926.

Buc'Hoz, P. J. *Recueil de secrets surs et experiments a l'usage des artistes*. 3 vols. Paris, 1783–1786.

Buck, Mitchell S. *Book Repair and Restoration*. Philadelphia, 1908.

The Carver and Gilder's Guide, and Frame Maker's Companion. London [c. 1870].

Chaptal, M.J.A. *Chemistry Applied to the Arts and Manufactures*. London, 1807.

———. *Elements of Chemistry*. Translated from the French. Philadelphia, 1801.

Clayton, Muriel. *The Print Collector*. New York, 1930.

Cooley, Arnold J. *A Cyclopedia of Practical Receipts and Collateral Information*. London, 1845.

Cumberland, George. *An Essay on the Utility of Collecting the Best Works of the Ancient Engravers of the Italian School* [1816]. London, 1827.

Cutbush, James. *The American Artists' Manual or Dictionary of Practical Knowledge in the Application of Philosophy to the Arts and Manufactures*. Philadelphia, 1814.

Davenport, Cyril. *Mezzotints*. London, 1904.

———. *Roger Paine: English Bookbinder of the 18th Century*. Chicago, 1929.

Dick, William. *Encyclopedia of Practical Receipts and Progresses*. New York, 1877.

An Expert. *The Book of Trade Secrets*. London, 1909.

Goss, Charles William Frederick. *Methods of Producing and Preserving Prints*. Aberdeen, 1915.

Gunn, Maurice James. *Print Restoration and Picture Cleaning*. London, 1911. 2d ed., 1922.

Haslum, W. *The Library Handbook of Genuine Trade Secrets and Instructions for Cleaning, Repairing, and Restoring Old Manuscripts, Engravings, and Books*. London, 1923.

Hecquet, Robert. *Catalogue des estampes gravées d'après Rubens. Auquel on a joint l'oeurve de Jordaens, & celle de Visscher. Avec un secret pour blanchir les estampes & en ôter les taches d'huile*. Paris, 1751.

Henley's Twentieth Century Book of Formulas, Processes and Trade Secrets. New York, 1913.

Jacobi, Charles T. *The Printers' Handbook*. London, 1891.

Jones, William A. *How to Make Home Happy*. London, 1852.

Lemery, Nicholas. *New Curiosities in Art and Nature*. London, 1711.

Leporini, Heinrich. *Der Kupferstichsammler*. Berlin, 1924.

Lonchamp, Frédéric-Charles. *Therapeutica graphica; ou, L'art de collectionner, de conserver et de restaurer les dessins, les manuscrits, les estampes et les livres*. Paris and Lausanne, 1930.

Lucanus, Friederich G. H. *Vollständige Anleitung zur Erhaltung, Reinigung und Wiederherstellung von Gemälden, Aquarellen, Kupferstichen, etc*. Halberstadt, 1812. 2d ed., Leipzig, 1828. 3d ed., 1846. 4th ed., 1887. Rev. ed., as *Die Praxis des Restaurators*, Halberstadt, 1929. 1929 ed., reprinted 1977.

Lumsden, E. S. *The Art of Etching*. Philadelphia, 1925.

Lydenburg, Harry M., and John Archer. *The Care and Repair of Books*. New York, 1931.

Maberly, Joseph. *The Print Collector*. London, 1844. Rev. ed., ed. Robert Hoe. London, 1880.

Mackenzie, Colin. *Five Thousand Receipts*. Philadelphia, 1829.

Martin, Jean Emile. *Art de dégrassier et de remettre a neuf le tissues*. Paris, 1828.

Mascall, Leonard. *A Profitable Booke*. London, 1588.

de Mayerne, Théodore Turquet. "Pictoria Sculptoria et Quæ Subalternarum Artium." British Museum, London, MS Sloane 2052.

———. *Le manuscrit de Turquet de Mayerne: Pictoria Sculptoria et Quæ Subalternarum Artium, 1620*. Lyon, 1974.

Merrifield, Mary P. *Original Treatises on the Arts of Painting*. London, 1849.

Meusnier, Georges. *De L'entretien et de la Restauration des tableaux, gravures, aquarelles et pastels*. Paris, 1909.

Musseau, J.C.L. *Manuel des amateurs d'estampes*. Paris, 1821.

The New Family Receipt Book. London, 1815.

Peacham, Henry. *The Compleat Gentleman*. London, 1661.

Plenderleith, H. J. *The Conservation of Prints, Drawings, and Manuscripts*. London, 1937.

Poortenaar, Jan. *Etkunst*. Amsterdam, 1930.

Porta, Giovanni Baptista della. *Natural Magick*. London, 1658.

Priestley, John. *A Familiar Introduction to the Theory and Practice of Perspective*. London, 1770.

Quicchelberg, Samuel van. *Inscriptiones vel tituli theatri amplissimi*. 1565.

Ruscelli, Girolamo. *The Secrets of Alexis of Piemont*. London, 1596.

Salmon, William. *Polygraphice*. London, 1673.

Schall, Joseph Frederick August. *Ausführliche Anleitung zur Restauration vergelbter, fleckiger und beschädigter Kupferstiche u.s.w*. Leipzig, 1863.

Schweidler, Max. *Die Instandsetzung von Kupferstichen, Zeichnungen, Büchern usw*. Stuttgart, 1938. 2d ed., 1949.

Seaman, Hervey J. *The Expert Cleaner*. London and New York, 1899.

Slater, George Herbert. *Engravings and Their Value*. London, 1891, rev. ed., 1912.

Spon, Ernest. *Workshop Receipts*. London, 1895.

Stannard, William J. *The Art Exemplar*. London, 1859.

The Techno-Chemical Receipt Book. Philadelphia, 1887.

Tothill, Richard. *The Arte of Limning*. London, 1573.

Tuer, Andrew W. *Bartolozzi and His Works*. London and New York, 1881.

Urbino, L. B., and Henry Day. *Art Recreations*. Boston, 1859. 2d ed., 1864. 3d ed., 1880.

Valuable Secrets in the Arts and Trades. Dublin, 1778.

White, John. *A Rich Cabinet with a Variety of Inventions*. London, 1653.

Whitman, Alfred. *The Print-Collector's Handbook*. London, 1901.

Willshire, William Hughes. *An Introduction to the Study and Collection of Ancient Prints*. London, 1877.

Wrinkles and Receipts. New York, 1875.

Contributors

Elmer Eusman completed his degree in conservation at Opleiding Restauratoren in Amsterdam in 1989. He has held internships at the Rijksmuseum, Amsterdam, the Boymans Museum, and the British Museum; and he served as a Mellon Fellow at the National Gallery of Art in the paper conservation department from 1990 to 1993. Eusman was a conservator at the Boymans Museum and the Library of Congress, and is now assistant paper conservator at the Baltimore Museum of Art.

Lisha Deming Glinsman joined the National Gallery of Art in 1987 as a conservation scientist; her work has included x-ray fluorescence analyses of Renaissance bronzes, ceramics, paintings, drawings, and, most recently, photographs. She received her B.A. in chemistry from the College of Saint Rose, Albany, and has worked in industry and as a research chemist for the National Institute of Standards and Technology. Her current research includes technical studies of Renaissance casting practices as a means of identifying forgeries, published in 1993 in *Conservation Research*. She has also published in *Archaeometry* (1993).

Susana M. Halpine received her master's degree from the Columbia College of Physicians and Surgeons in 1987. A Mellon Fellow from 1988 to 1990, she is now a biochemist in the scientific research department at the National Gallery of Art. Prior to her fellowship, she worked as a staff associate in the Protein Analysis Facility of the Howard Hughes Medical Institute at Columbia University. She has used high-performance liquid chromatography extensively for amino acid analysis, protein sequencing, and fingerprinting, as well as for purifying synthetic DNA and peptides.

Constance McCabe, photograph conservator at the National Gallery of Art, earned her M.F.A. in 1982 at Rochester Institute of Technology in the School of Photographic Arts and Sciences. She has served as a conservation intern with the International Museum of Photography at George Eastman House and as a research assistant at the Image Permanence Institute, both in Rochester. Following six years as photograph conservator at the National Archives and Records Administration, she joined the National Gallery of Art in 1991.

Albert Marshall received his undergraduate degree in art from Kenyon College and his M.S. in art conservation from the University of Delaware. He has held internships at Olin Conservation, Great Falls, Virginia; the Joseph Cornell Study Center, National Museum of American Art, Smithsonian Institution; and the objects conservation department, National Gallery of Art. He was a National Gallery of Art Mellon Fellow from 1989 to 1991 and an objects conservator for the systematic catalogue from 1991 to 1993. At present he is the conservator at Hillwood Museum, Washington, D.C.

Mark Stevenson received his B.A. in drawing and printmaking at Southern Illinois University and his M.A. from the Buffalo State College Art Conservation Program. His certificate in the conservation of works of art is from the Fogg Museum, Harvard University. From 1988 to 1990 he was a Mellon Fellow at the National Gallery of Art. He has also worked in the drawing and print department, Indianapolis Museum of Art, and was assistant conservator of paper at the Nelson-Atkins Museum from 1990 to 1993. He is an academic lecturer and a paper conservator in private practice in the Kansas City area.

Studies in the History of Art
Published by the National Gallery of Art, Washington

This series includes: Studies in the History of Art, collected papers on objects in the Gallery's collections and other art-historical studies (formerly *Report and Studies in the History of Art*); Monograph Series I, a catalogue of stained glass in the United States; Monograph Series II, on conservation topics; and Symposium Papers (formerly Symposium Series), the proceedings of symposia sponsored by the Center for Advanced Study in the Visual Arts at the National Gallery of Art.

[1] *Report and Studies in the History of Art,* 1967

[2] *Report and Studies in the History of Art,* 1968

[3] *Report and Studies in the History of Art,* 1969 [In 1970 the National Gallery of Art's annual report became a separate publication.]

[4] *Studies in the History of Art,* 1972

[5] *Studies in the History of Art,* 1973 [The first five volumes are unnumbered.]

6 *Studies in the History of Art,* 1974

7 *Studies in the History of Art,* 1975

8 *Studies in the History of Art,* 1978

9 *Studies in the History of Art,* 1980

10 *Macedonia and Greece in Late Classical and Early Hellenistic Times,* edited by Beryl Barr-Sharrar and Eugene N. Borza. Symposium Series I, 1982

11 *Figures of Thought: El Greco as Interpreter of History, Tradition, and Ideas,* edited by Jonathan Brown, 1982

12 *Studies in the History of Art,* 1982

13 *El Greco: Italy and Spain,* edited by Jonathan Brown and José Manuel Pita Andrade. Symposium Series II, 1984

14 *Claude Lorrain, 1600–1682: A Symposium,* edited by Pamela Askew. Symposium Series III, 1984

15 *Stained Glass before 1700 in American Collections: New England and New York (Corpus Vitrearum Checklist I),* compiled by Madeline H. Caviness et al. Monograph Series I, 1985

16 *Pictorial Narrative in Antiquity and the Middle Ages,* edited by Herbert L. Kessler and Marianna Shreve Simpson. Symposium Series IV, 1985

17 *Raphael before Rome,* edited by James Beck. Symposium Series V, 1986

18 *Studies in the History of Art,* 1985

19 *James McNeill Whistler: A Reexamination,* edited by Ruth E. Fine. Symposium Papers VI, 1987

20 *Retaining the Original: Multiple Originals, Copies, and Reproductions.* Symposium Papers VII, 1989

21 *Italian Medals,* edited by J. Graham Pollard. Symposium Papers VIII, 1987

22 *Italian Plaquettes,* edited by Alison Luchs. Symposium Papers IX, 1989

23 *Stained Glass before 1700 in American Collections: Mid-Atlantic and Southeastern Seaboard States (Corpus Vitrearum Checklist II),* compiled by Madeline H. Caviness et al. Monograph Series I, 1987

24 *Studies in the History of Art,* 1990

25 *The Fashioning and Functioning of the British Country House,* edited by Gervase Jackson-Stops et al. Symposium Papers X, 1989

26 *Winslow Homer,* edited by Nicolai Cikovsky, Jr. Symposium Papers XI, 1990

27 *Cultural Differentiation and Cultural Identity in the Visual Arts,* edited by Susan J. Barnes and Walter S. Melion. Symposium Papers XII, 1989

28 *Stained Glass before 1700 in American Collections: Midwestern and Western States (Corpus Vitrearum Checklist III),* compiled by Madeline H. Caviness et al. Monograph Series I, 1989

29 *Nationalism in the Visual Arts,* edited by Richard A. Etlin. Symposium Papers XIII, 1991

30 *The Mall in Washington, 1791–1991,* edited by Richard Longstreth. Symposium Papers XIV, 1991

31 *Urban Form and Meaning in South Asia: The Shaping of Cities from Prehistoric to Precolonial Times,* edited by Howard Spodek and Doris Meth Srinivasan. Symposium Papers XV, 1993

32 *New Perspectives in Early Greek Art,* edited by Diana Buitron-Oliver. Symposium Papers XVI, 1991

33 *Michelangelo Drawings,* edited by Craig Hugh Smyth. Symposium Papers XVII, 1992

34 *Art and Power in Seventeenth-Century Sweden,* edited by Michael Conforti and Michael Metcalf. Symposium Papers XVIII *(withdrawn)*

35 *The Architectural Historian in America,* edited by Elisabeth Blair MacDougall. Symposium Papers XIX, 1990

36 *The Pastoral Landscape,* edited by John Dixon Hunt. Symposium Papers XX, 1992

37 *American Art around 1900,* edited by Doreen Bolger and Nicolai Cikovsky, Jr. Symposium Papers XXI, 1990

38 *The Artist's Workshop,* edited by Peter M. Lukehart. Symposium Papers XXII, 1993

39 *Stained Glass before 1700 in American Collections: Silver-Stained Roundels and Unipartite Panels (Corpus Vitrearum Checklist IV),* compiled by Timothy B. Husband. Monograph Series I, 1991

40 The Feast of the Gods: *Conservation, Examination, and Interpretation,* by David Bull and Joyce Plesters. Monograph Series II, 1990

41 *Conservation Research.* Monograph Series II, 1993

42 *Conservation Research: Studies of Fifteenth- to Nineteenth-Century Tapestry,* edited by Lotus Stack. Monograph Series II, 1993

43 Eius Virtutis Studiosi: *Classical and Postclassical Studies in Memory of Frank Edward Brown,* edited by Russell T. Scott and Ann Reynolds Scott. Symposium Papers XXIII, 1993

44 *Intellectual Life at the Court of Frederick II Hohenstaufen,* edited by William Tronzo, Symposium Papers XXIV, 1994

45 *Titian 500,* edited by Joseph Manca. Symposium Papers XXV, 1994

46 *Van Dyck 350,* edited by Susan J. Barnes and Arthur K. Wheelock, Jr. Symposium Papers XXVI, 1994

47 *The Formation of National Collections of Art and Archaeology,* edited by Gwendolyn Wright. Symposium Papers XXVII*

48 *Piero della Francesca and His Legacy,* edited by Marilyn Aronberg Lavin. Symposium Papers XXVIII, 1995

49 *The Interpretation of Architectural Sculpture in Greece and Rome,* edited by Diana Buitron-Oliver. Symposium Papers XXIX*

50 *Federal Buildings in Context: The Role of Design Review,* edited by J. Carter Brown. Symposium Papers XXX*

51 *Conservation Research 1995.* Monograph Series II, 1995

52 *Saint-Porchaire Ceramics,* edited by Shelley Sturman and Daphne Barbour*

53 *Imagining Modern German Culture, 1889–1910,* edited by Françoise Forster-Hahn. Symposium Papers XXXI*

* Forthcoming